AF851024

# Nuclear Physical Methods in Radioecological Investigations of Nuclear Test Sites

# NATO Science Series

*A Series presenting the results of activities sponsored by the NATO Science Committee. The Series is published by IOS Press and Kluwer Academic Publishers, in conjunction with the NATO Scientific Affairs Division.*

| | |
|---|---|
| **A. Life Sciences** | IOS Press |
| **B. Physics** | Kluwer Academic Publishers |
| **C. Mathematical and Physical Sciences** | Kluwer Academic Publishers |
| **D. Behavioural and Social Sciences** | Kluwer Academic Publishers |
| **E. Applied Sciences** | Kluwer Academic Publishers |
| **F. Computer and Systems Sciences** | IOS Press |

| | |
|---|---|
| **1. Disarmament Technologies** | Kluwer Academic Publishers |
| **2. Environmental Security** | Kluwer Academic Publishers |
| **3. High Technology** | Kluwer Academic Publishers |
| **4. Science and Technology Policy** | IOS Press |
| **5. Computer Networking** | IOS Press |

**NATO-PCO-DATABASE**

The NATO Science Series continues the series of books published formerly in the NATO ASI Series. An electronic index to the NATO ASI Series provides full bibliographical references (with keywords and/or abstracts) to more than 50000 contributions from international scientists published in all sections of the NATO ASI Series.
Access to the NATO-PCO-DATA BASE is possible via CD-ROM "NATO-PCO-DATA BASE" with user-friendly retrieval software in English, French and German (WTV GmbH and DATAWARE Technologies Inc. 1989).

The CD-ROM of the NATO ASI Series can be ordered from: PCO, Overijse, Belgium

**Series 1: Disarmament Technologies – Vol. 31**

# Nuclear Physical Methods in Radioecological Investigations of Nuclear Test Sites

edited by

Siegfried S. Hecker
Caroline F. V. Mason
Los Alamos National Laboratory,
Los Alamos,
New Mexico, U.S.A.

and

Kairat K. Kadyrzhanov
Serge B. Kislitsin
Institute of Nuclear Physics,
National Nuclear Centre of the Republic of Kazakhstan,
Almaty, Republic of Kazakhstan

**Kluwer Academic Publishers**

Dordrecht / Boston / London

Published in cooperation with NATO Scientific Affairs Division

Proceedings of the NATO Advanced Research Workshop on
Nuclear Physical Methods in Radioecological Investigations of Nuclear Test Sites
Almaty, Kazakhstan
7–10 June 1999

A C.I.P. Catalogue record for this book is available from the Library of Congress.

ISBN 0-7923-6447-3

---

Published by Kluwer Academic Publishers,
P.O. Box 17, 3300 AA Dordrecht, The Netherlands.

Sold and distributed in North, Central and South America
by Kluwer Academic Publishers,
101 Philip Drive, Norwell, MA 02061, U.S.A.

In all other countries, sold and distributed
by Kluwer Academic Publishers,
P.O. Box 322, 3300 AH Dordrecht, The Netherlands.

*Printed on acid-free paper*

All Rights Reserved
© 2000 Kluwer Academic Publishers
No part of the material protected by this copyright notice may be reproduced or utilized in any form or by any means, electronic or mechanical, including photocopying, recording or by any information storage and retrieval system, without written permission from the copyright owner.

Printed in the Netherlands.

## TABLE OF CONTENTS

## METHODS OF RETROSPECTIVE ESTIMATION OF THE ABSORBED DOSE IN ENVIRONMENTAL OBJECTS AND THE HUMAN BODY



## MODERN METHODOLOGY AND EQUIPMENT FOR FIELD RADIOMETRY AND SPECTROMETRY

# PREFACE

The end of the Cold War created a new wave of co-operation between former adversaries. The need for collaboration and understanding is central to post-Cold War policy as a way to lessen tension between former adversaries. A new scientific dialogue, for example, is reducing apprehension through increased understanding and is providing a mechanism for scientists of different nationalities and diverse disciplines to discuss universal challenges.

In addition, the end of the Cold War has also brought to light the environmental hazards remaining at many nuclear test sites around the world. The long-term ecological problems resulting from nuclear testing are only now receiving proper attention as are the long-term potential health risks. A significant effort is now being focused on the principal nuclear test site of the Former Soviet Union, the Semipalatinsk Nuclear Test Site (SNTS) in Kazakhstan, where over 400 nuclear tests were conducted.

Within this framework, the Advanced Research Workshop "Nuclear Physical Methods in Radioecological Investigations of Nuclear Test Sites," under the sponsorship of NATO, was planned and implemented by the National Nuclear Centre of the Republic of Kazakhstan and the Department of Energy's Los Alamos National Laboratory, USA.

The goals of this workshop were to evaluate existing methods for determining artificial radionuclides and to look for more effective alternative methods, to focus on gaps in information, to produce a path forward in evaluating ecological priorities at such sites, and to consider methodologies for determining risk.

Radioecological investigations of nuclear test sites are undertaken for many purposes, with different methods being used according to the goals of the investigative work. However, all of these investigations are based upon information about radionuclide concentrations in the environment. Reliable and accurate data are crucial if the work is to be carried out properly. Many different methods are used to determine radionuclide concentrations, (e.g., plutonium alpha spectrometry, gamma spectrometry, etc.). Various radiochemical techniques are used for plutonium separation, alpha spectrometric source preparation, and americium and strontium-90 determination. Each of these techniques has advantages and disadvantages. Unfortunately, the effectiveness of each method is calculated infrequently and is based on literature data only.

Investigations of the nuclear test sites indicate that radionuclides exist at some locations at levels much higher than those accounted for by global fallout. To quantify this finding, the application of gamma-spectrometry for plutonium, for example, could be an effective and innovative tool rather than the normally used alpha-spectrometry. Alternative methods could also be proposed for determination of other radionuclides such as americium, strontium-90, etc.

The Advanced Research Workshop, "Nuclear Physical Methods in Radioecological Investigations of Nuclear Test Sites," took place in the conference rooms of the Institute of Nuclear Physics of the National Nuclear Centre of the Republic of Kazakhstan, Almaty, Kazakhstan, June 7- 10, 1999. Representatives from eight countries as well as the International Atomic Energy Agency were present. The thirty-one papers presented covered topics that included dose assessment, risk, speciation and transport of radionuclides, measurement and separation of radionuclides, sampling, both aerial and ground sampling, and remedial options. The nuclear test sites discussed included the SNTS, the Nevada Test Site, the Pacific atolls, and the Maralinga Test Site in Australia. In addition, other release sites that were discussed included Chelyabinsk, Chernobyl, Palomares (Spain), and Los Alamos (USA). Potential release sites such as those in the Arctic were also considered. Attendance at the main oral sessions averaged about 60 attendees. Many scientists from the host institution chose to listen to the talks. The simultaneous translation, Russian to English and vice versa, allowed all attendees to participate. There was a lively final session when the participants discussed a path forward towards solving issues surrounding rehabilitation of nuclear test sites for productive land use (farming, mining etc).

The Recommendations of the Conference follow.

**Scientific Recommendations**

1. Put in perspective the risk associated with past radionuclide contamination as compared to other risks and costs. Other risks tend to be greater than those associated with past radionuclide contamination.
2. Develop rapid *in-situ* methods of radionuclide determination, especially for plutonium. These faster methods will speed up contaminant characterisation and thereby lower cost.
3. Use an interdisciplinary approach to solving problems so that many disciplines are involved in an assessment of risk from radionuclides.
4. Promote a better understanding of heterogeneous contamination on dose and how to incorporate this information into dose assessment. How "hot spots" are handled creates large differences in risk assessments.
5. Encourage more consistent use of the International System of Units (SI) in reporting results. At the conference, different speakers used different measurements, for example Bq/Kg and pCi/g, thereby making direct comparisons cumbersome.

**International Recommendations**

1. Continued collaborations are particularly helpful for Kazakhstanto keep it from being isolated from the mainstream of scientific discoveries in this field.

2. Many of the problems discussed at the Workshop are experienced world-wide and need to be approached in the same way in all areas, that is by using unified methods of reporting data and unified methods of calculating risk.
3. Increased availability of former Soviet military data on the SNTS would be extremely helpful to the Republic of Kazakhstan.
4. The Republic of Kazakhstan will play a more active role in assessing risk at the SNTS. United Nations Resolution-A/53/424 calls for international co-operation and co-ordination of the human and ecological rehabilitation and economic development of the Semipalatinsk region of Kazakhstan. This resolution is helpful. A comprehensive study of the dose and risk from past exposure is necessary.
5. A comprehensive risk assessment for the entire Republic of Kazakhstan would provide useful data, as there are many other sources of contamination outside the SNTS (e.g. uranium mining and milling, nuclear tests, etc.)
6. There is a need to accept the public's desire (and even demand) for more information about contamination at SNTS. Contamination is a very emotional issue that must be discussed with the public. For example, Ryspek Ibraev, deputy leader of the "Nevada Semipalatinsk Movement," attended all the sessions and asked many questions. His main concern was the perception that the data were not being freely disseminated.
7. The Conference supports United Nations Resolution-A/53/424 on the Semipalatinsk Test Site. This resolution calls on the world community to participate in solving the problems arising from nuclear tests by introducing rehabilitation measures to the territory of the test site. The Conference calls on NATO to take part in fulfilling such rehabilitation and research projects. The Republic of Kazakhstan will play a more active role in assessing risk at the SNTS.
8. The Conference recommends that the United Nations include the National Nuclear Centre among the initiators of the UN programme on the research and implementation of rehabilitation measures at the SNTS.
9. The Conference recognises the advanced scientific accomplishments of Kazakhstan organisations, especially the National Nuclear Centre, in engaging the problems of the SNTS.

In his closing comments, Conference Co-Chairman, Dr. Siegfried Hecker, acknowledged his co-chairman, Director of the National Nuclear Centre, Dr. Yuri Cherepnin, who was unable to attend owing to illness. In addition, Dr. Hecker thanked the sponsors, especially NATO, the organisers (in particular Dr. Serge Kislitsin of NNC and Dr. Caroline Mason of Los Alamos National Laboratory), the interpreters at the NNC, the participants who made the conference such a success, and the audience for their sustained presence and participation in lively discussions.

Caroline Mason
Los Alamos National Laboratory
Los Alamos, New Mexico, USA

**ACKNOWLEDGEMENTS**

This Advanced Research Workshop that took place in Almaty, Kazakhstan from June 7 to 10, 1999 generated important interactions among scientists from diverse nations and provided a mechanism for scientists of varied disciplines to discuss challenges that confront many countries around the world. This volume provides a strong foundation on which to build further multilateral programmes that can enhance understanding and cooperation in addressing the issues surrounding the sites where nuclear weapons were once tested.

The authors are grateful for the financial support from the North Atlantic Treaty Organization (NATO) through the Disarmament Program Advanced Research Workshop (DISRM-ARW-973310), with special thanks to Nancy Schulte of NATO. Also we acknowledge financial support from the Los Alamos National Laboratory, USA, and the Institute of Nuclear Physics, National Nuclear Centre of the Republic of Kazakhstan. Without such support, this ARW and the exchanges among the workshop participants would not have been possible.

The members of the organizing committee would like to express their gratitude to the speakers from Denmark, Hungary, Kazakhstan, Norway, Russia, Slovakia, Ukraine, and the USA, as well as to the participants from the IAEA for developing their technical papers and presentations. In addition the committee would like to extend its thanks to everyone who attended the ARW and contributed to the discussions.

The Institute of Nuclear Physics near Almaty provided excellent meeting facilities and arrangements. The staff of the Institute of Nuclear Physics, which handled all the logistical arrangements for the meeting, transportation, and lodging in Kazakhstan, made major contributions to the success of the ARW.

The ARW presentations and discussions were enhanced by the superb quality of the professional interpreters Tatjana Bolgova, Julie Logvinova, and Ludmila Zaitseva whose work was outstanding.

Finally, the authors gratefully acknowledge the extensive work of writer-editor, Elizabeth Katz and compositor, Jan Dye of the Los Alamos National Laboratory for producing a manuscript of high quality appropriate to the importance of the ARW that this book represents.

# WELCOME AND INTRODUCTION

## GREETINGS FROM THE NATIONAL NUCLEAR CENTRE OF THE REPUBLIC OF KAZAKHSTAN

The National Nuclear Centre of the Republic of Kazakhstan and the Institute of Nuclear Physics extend greetings to all the participants of the North Atlantic Treaty Organizaton Advanced Research Workshop, "Nuclear Physical Methods in Radioecological Investigations of Nuclear Test Sites," taking place in Almaty. It is a great honour for us to host this workshop and to have our foreign visitors here, especially at the same time as our biennial international conference, "Nuclear and Radiation Physics."

It is not by chance that our Centre hosts this symposium devoted to the challenges of former nuclear test sites. You know that the Semipalatinsk Nuclear Test Site here in Kazakhstan was in continuous operation for forty years. We are truly privileged to have experts gathered from many countries to openly discuss the challenges related to radioecological issues.

We particularly welcome our Co-Chairman, Dr. Siegfried Hecker, from Los Alamos National Laboratory in the United States, and all the participants who have travelled to Kazakhstan to contribute their experience to this conference.

This conference will surely advance international cooperation in this important area. We wish you all a fruitful and interesting conference.

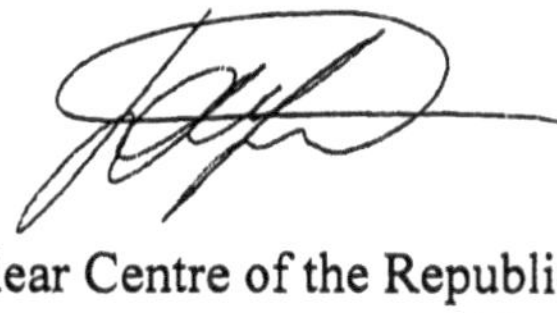

Yuri Cherepnin
June 1999
Director, National Nuclear Centre of the Republic of Kazakhstan
Almaty, Kazakhstan

# OVERVIEWS AND PERSPECTIVES

# A PATH TO A NUCLEAR FUTURE

S. S. HECKER
*Senior Fellow, Los Alamos National Laboratory*
*Los Alamos, NM 87545 USA*

## 1. Introduction

It is a great honour for me to share the stage during the opening of this conference and the NATO workshop with my colleagues from Kazakhstan and Russia—and with the participants from nations around the world. This conference and the workshop deal with nuclear topics ranging from science to social and political issues. As Director of the Los Alamos National Laboratory from 1986 through 1997, I had the opportunity to witness enormous changes in the nuclear world – and, for that matter, enormous changes in world affairs. Since leaving the directorship, I have focused most of my efforts on nuclear issues, especially on the importance of international collaboration in securing a beneficial nuclear future. I have spent much time in Russia over the past seven years, and this is my second visit to Kazakhstan. I have also been to the United Kingdom, France, and China to discuss these issues. In the spirit of introducing both the conference and the workshop, I would like to share my thoughts on how to build a beneficial nuclear future.

## 2. Remarks by Yuli Borisovich Khariton

I find it most appropriate to begin my talk with a quote from Academician Yuli Borisovich Khariton, who led the Soviet nuclear weapons programme for almost its entirety. In 1995, Academician Khariton contributed a written paper for the Oppenheimer lecture at Los Alamos. I quote from the closing paragraph of his very thoughtful lecture.

Conscious of my participation in remarkable scientific and engineering achievements, which led to mankind's possession of a practically inexhaustible source of energy, today at a more mature age, I am no longer sure whether mankind has matured enough to possess this energy. I realise our participation in the terrible death of people and in the dreadful damage inflicted upon the nature of our home, the Earth. Words of repentance can certainly change nothing. Please God, let's hope that those



*S.S. Hecker et al. (eds.), Nuclear Physical Methods in Radioecological Investigations of Nuclear Test Sites*, 3-9.
© 2000 *Kluwer Academic Publishers. Printed in the Netherlands.*

who come after us will find the ways, find in themselves the firmness of spirit and determination, in striving for the best, not to do the worst.

I believe that with these words Academician Khariton laid out the challenge for a path to a beneficial nuclear future. In doing so, he mentioned the following three principal issues:

- The promise of practically inexhaustible energy
- The question of mankind's maturity to possess such destructive power
- The concern about damage to the Earth.

I will develop my remarks around Academician Khariton's themes and then explain my perspective of what needs to be done to accomplish the best, not the worst of these issues. Also, I will try to connect my remarks to the specific conference and workshop objectives and to the nuclear issues here in Kazakhstan.

### 3. Nuclear Energy—Remarkable Achievement and Promise

As human aspirations develop, so grows the demand for energy use. Energy is a substitute for human labour. It is the key to improved living for all the peoples of the world.

Nuclear energy holds a special promise of an abundant, clean, affordable, and almost inexhaustible source of energy. My colleagues from the various installations of the Russian Ministry of Atomic Energy (Minatom) believe in this promise with their hearts. My April 1998 visit with the scientists and engineers of the Kazakhstan National Nuclear Centre (NNC) here in Almaty and at the Semipalatinsk Test Site convinced me that my colleagues here at the Kazakhstan NNC believe this. Hence, I do not need to impress upon our hosts the importance of nuclear energy.

Nevertheless, allow me to give you a brief historical perspective on nuclear energy – and then add some of the realities as we approach the next century. Shortly after the discovery of natural radioactivity 100 years ago, scientists such as Frederick Soddy, a colleague of Ernest Rutherford's at Cambridge, United Kingdom, began to dream about "radioactivity's" pointing to inexhaustible power. By 1903, writers in the popular press (such as Gustave Le Bon in France) said that soon "the poor will be equal to the rich and there will be no more social problems" because of the powers of radioactivity. Just to remind you, these dreams existed 35 years before the discovery of fission by Hahn and Strassmann in Germany. One year later in 1938 in Russia, Khariton and Zel'dovich wrote their first paper on the possibility of a chain reaction, and four years later in Chicago on December 2, 1942, Enrico Fermi demonstrated the first nuclear chain reaction.

Although the immediate focus of these efforts was to build an atomic bomb, the dream of almost inexhaustible energy was present from the beginning. Today 400 nuclear power plants supply a substantial amount of electricity (and heat) around the world. However, the construction of nuclear power plants has slowed considerably because of a number of factors—some inherent to nuclear power, others simply because of economic considerations.

The dream of the breeder reactor—the potential of breeding plutonium from fertile uranium fuel or the potential of using a thorium-232 nuclear fuel cycle to breed uranium-233—remains but a dream. The United States considers that breeder reactors will not be economically attractive until the last half of the next century. France has slowed down its breeder reactor programme—now considering it for 20 to 30 years in the future. Japan continues to explore the breeder option, and Russia is ready to build the BN-800 breeder reactor, but simply does not have the money.

Nevertheless, thermal nuclear reactors are being considered in many parts of the developing world. As one looks at the overall energy projections over the next 100 years, it appears inevitable that nuclear energy must play a substantial role. The prospects for a revival of nuclear energy are substantially increased by current concerns over the ecological damage to the Earth from the burning of fossil fuels. However, to have nuclear energy play such an important role for the future of mankind, we must first realise that dark shadows are currently cast on the potentially bright future of nuclear energy. We must not only recognise these shadows, but we must collectively deal with them.

The shadows to which I refer are the ones expressed by Academician Khariton—the question of mankind's maturity to possess nuclear weapons and the concern about the legacy to Mother Earth.

## 4. Nuclear Weapons—Deterrence or Destruction?

Academician Khariton expressed his concern about mankind's maturity to possess such a destructive force. In this century, we have acquired the technological means of our self-destruction before we have learned how to live together peacefully. In 1995, Khariton must have reflected on the fact that the rivals in the Cold War built nearly 100 000 nuclear weapons. One of the bombs designed by the Soviets had a yield of 100 megatons (5000 times the destructive power of the bombs at Hiroshima and Nagasaki). No wonder many people around the world are concerned about a nuclear confrontation that could destroy the world as we have known it.

However, this incredibly destructive power made mankind pause because it recognised its own collective mortality. Although until the end of World War II, instruments of war were continuously perfected to annihilate each other, this maddening race appears to have been halted in 1945. The atomic bombs at Hiroshima and Nagasaki changed the world forever. The Cold War standoff, in fact, did work. It prevented World War III. However, with tens of thousands of nuclear weapons ready to be delivered in minutes to hours, it was at best an uneasy peace.

It is interesting to look back in history to see how the public's fear of things nuclear has grown. The first warnings of the dangers of "radioactivity" emerged simultaneously with the great hope for inexhaustible power. The French writer, Gustave LeBon by imagining a radioactive device warned in 1903 about "blowing up the Earth" at the touch of a button. In his 1913 novel, *The World Set Free*, the British writer, H.G. Wells created the idea of a radioactive chain reaction that could lead to "atomic bombs." So, the public already had been conditioned for the awesome destructive power of the atomic bombs at Hiroshima and Nagasaki. To this date, I believe, the public has continued to associate the benefits of nuclear energy with the dark side of "blowing up the world."

For nations, of course, it was a very different matter. The United States began the Manhattan Project to build the atomic bomb in a race against Hitler's Germany. The United States opened its arms to the great European scientists who fled Hitler's persecution and joined them with American scientists to build the first bombs in time to end the war with Japan. The Soviet Union did not mount a serious effort to build the bomb during the war because it had to mobilise all of its resources to fight Hitler. It is now clear, however, that the knowledge to build such a device existed.

As Academician Khariton told me, it was no surprise for many nations to develop the basic knowledge on how to build a bomb since all of the scientists at the time had the same basic roots—the great school of European physics. In 1992 Khariton also told me the story of the first Soviet bomb that was exploded here at the Semipalatinsk Test Site almost 50 years ago. He told me openly that the information Soviet intelligence agents received from Klaus Fuchs (at Los Alamos with the British team of scientists) helped to shorten the time required for developing the first Soviet bomb.

Once Stalin heard about the devastation in Japan, he was not to be denied his own atomic bombs. That is not surprising. However, the reaction of the scientists at that time is also very interesting. Khariton told me that he believed it was imperative that the United States would not have a monopoly on such a powerful weapon. He and others worked with great patriotism for the freedom of their country. Andrei Sakharov in his memoirs added, "... I regarded myself as a soldier in this new scientific war." He continued, "But above all, I felt myself committed to the goal which I assumed was Stalin's as well: after a devastating war, to make the country strong enough to ensure peace."

General Charles De Gaulle in 1958 spoke on behalf of the French nuclear weapons programme as it came into existence. He noted, "France, by acquiring nuclear arms is performing a service for world equilibrium." I mention these examples to make the point that a nation's drive to acquire nuclear weapons may be undeterred. We, of course, know that in addition to the U.S.A., the Soviet Union, and France, the United Kingdom, and China acquired their own nuclear weapons. As already mentioned, the resulting Cold War standoff did not lead to nuclear war. However, there has always been great concern about potential instabilities should additional nations acquire nuclear weapons. That is the reason that the 1968 Nonproliferation Treaty (NPT) was so important and why its extension in 1995 was crucial.

Over the years, some nations started nuclear weapons programmes and then abandoned them. Sweden, South Africa, Argentina, and Brazil took different paths, but all lead to the same end point. Other nations developed nuclear weapons clandestinely and still have them. Israel, India, and Pakistan fall into that category. And, as we now know, India and Pakistan decided to test one year ago and declared their nuclear capabilities. Still other nations continue to attempt to acquire nuclear weapons by any means. At this time the world is concerned about the actions of Iraq, Iran, and North Korea in the nuclear weapons arena.

One of the most important decisions made to contain the spread of nuclear weapons was the one made right here in Kazakhstan. Your country along with the Ukraine and Belarus became an instant nuclear power when the Soviet Union dissolved in December 1991. With the urging of the United States and Russia, you decided to give these weapons back to Russia and to sign the NPT. So did the Ukraine and Belarus. These actions were critical to containing the potential instabilities in the nuclear affairs of nations. The challenge as I see it now is for the United States and Russia to dramatically reduce their nuclear arsenals, for other nuclear weapons states to show restraint, and for the rest of the world to accept the NPT principle—that there shall be no new nuclear weapons states. Such an action is the best way to avoid reliving the nightmares that plagued so many during the darkest days of the Cold War.

**5. Nuclear Proliferation—Theft or Diversion of Weapons or Materials?**

Whereas the end of the Cold War dramatically reduced the potential of a global nuclear confrontation, it actually increased the possibility that nuclear weapons or nuclear materials could be diverted or stolen. The greatest concern is in the transition of the former Soviet Union to a democratic Russia and to other democratic republics. During the Cold War, the Soviet Union had an excellent record of protecting its nuclear weapons, its nuclear materials, and its nuclear secrets. The centrally controlled system of physical and personnel security combined with the closed nature of the Soviet Union protected nuclear matter well.

However, on Dec. 25, 1991, President Gorbachev dissolved the Soviet Union. The system of central control was replaced by democratic governments and by free-market economic systems. Moreover, some of the nuclear materials were no longer in Russia. For example, Kazakhstan not only inherited nuclear weapons (which it then returned) but also nuclear reactors here in Almaty, at Aktau, and at the Semipalatinsk Test Site. The economic hardship incurred during the transition from previous political and economic systems to new ones has put enormous stress on the nuclear enterprises of Russia and the other Newly Independent States (NIS).

I have tried to convince the U.S. government that it is in its own interest to help Russia and the other NIS in making this transition. The nuclear materials protection, control, and accounting (MPC&A) programme (a joint endeavour between the U.S.A. and Russia, Kazakhstan, etc.) has made significant progress in changing an old system to one that meets the new threats—open borders, less central control, organised crime, stressed citizens, and insufficient money. I saw the new systems in place in Kazakhstan last year during my visit here. However, I fully recognise that the very good record that Russia and the other states have had since 1992 is mostly the result of the enormous dedication and loyalty of the people working in the nuclear enterprises of Russia, Kazakhstan, and the other states.

Unfortunately, if people are stressed for so long, eventually any system will fail. That is the reason it is imperative that other industrialised nations work closely with Russia and the other states to help them make this difficult transition. These are vulnerable times. We must work together to overcome that vulnerability. Programmes such as the MPC&A, the International Science and Technology Centre (ISTC), the Nuclear Cities Initiative, the HEU (highly-enriched uranium) purchase, and plutonium disposition are all very important. Just think. If sufficient material is stolen or diverted from any of the installations of Russia or Kazakhstan (or from the other countries), and a nuclear bomb is produced (unfortunately, it is frightfully easy to build a crude bomb once the material is available) and detonated anywhere in the world—not only will the toll of human suffering be enormous, but the future of nuclear power will be set back for a long, long time.

## 6. Nuclear Legacy—"Damage to the Earth"

Academician Khariton wrote about the damage that the nuclear enterprise has inflicted upon our home, the Earth. I am sure that he was concerned that over the years the extraction of uranium as the starting material, the processing of these materials, the production of plutonium in reactors, the reprocessing of spent fuel into weapons-grade material, the testing of nuclear devices (especially those tested in the atmosphere), the decommissioning of nuclear reactors (including those used in the naval sector), and the handling and eventual disposition of radioactive wastes have all stressed the environment we live in. And, of course, he was right. Much like any industrial venture, the nuclear enterprise had its negative side effects on the ecology of the Earth. Much of the work done on the military side was done in haste—driven by the enormous pressures of the arms race between the U.S.A. and the Soviet Union.

Consequently, the U.S.A. and Russia (and some other NIS) now face an enormous job of cleaning up the legacy of the nuclear enterprises. Many people have been highly critical of the environmental records of both the U.S.A. and the Soviet Union during the Cold War. However, others have viewed these sites as the battlefields of the Cold War. And, although these battlefields are radioactively contaminated (in some cases, highly so), they are battlefields that produced few casualties compared to the battlefields of the great wars that preceded the Cold War. The ecological issues are naturally issues of great emotion. These are issues that are local. All citizens who live around these battlefields have concerns about the health of their children. The locations of these battlefields were, of course, the result of national decisions made by national governments.

Nowhere is this issue of greater concern than right here in Kazakhstan, where it was the action of a government (the Soviet government) that no longer exists that made the decisions leading to the placements of some of its battlefields in what is now your independent nation. One of these concerns, the radioecological issues associated with nuclear test sites, is, of course, the subject of the NATO workshop being conducted here this week.

I believe we must now pool our resources, both intellectual and financial, to help clean up those ecological problems that cause a concern for our children now as well as in the future. This is a time for action and co-operation, not for finger pointing. I believe that because of the financial hardships in Russia and the other NIS, the industrialised nations must step forward now to help out. These nations cannot be expected to clean up the problems caused by the Soviet Union, but they can help to launch the assessments and the initial efforts in the right direction. I believe that in all countries cleaning up the legacy of the nuclear programmes—be these programmes of a military or civilian origin—is a required step before the public will support a major expansion of peaceful nuclear power production. The ecological shadow of this legacy on the future of nuclear power must be erased together.

One additional shadow is the concern over nuclear safety. Again, I refer both to military and civilian power systems. As Vladimir Belugin, former director of VNIIEF/Arzamas-16, once told me, "Anyone's nuclear accident is everyone's nuclear accident." There is much that has been done to answer the questions raised by the Chernobyl accident, and there is much more that can be done by the technical community to provide future nuclear reactors with greater intrinsic safety. Likewise, all states that have nuclear weapons must place the safety of these weapons first and foremost on the national agenda. That is why the nuclear weapons stewardship programmes of the U.S.A. and Russia are so important today.

## 7. Scientific Co-operation—Path to a Nuclear Future

To come to my last point, I would like to stress that co-operation among the scientists of the world will lead us to the desirable nuclear future that Academician Khariton, J. Robert Oppenheimer, Enrico Fermi, and so many others envisioned. To exploit the practically inexhaustible source of energy of the nucleus, we must work collectively to remove the dark shadows cast upon that bright future. Together we must work to prevent the proliferation of nuclear weapons. Together we must work to make nuclear reactors safer. And, together we must work to deal with the ecological legacy of the military and civilian nuclear programmes. Otherwise, the public will not accept our premise that nuclear energy is the preferred path to better living.

Once again, the U.S.A. and Russia have a great opportunity to lead the way. Not only can they demonstrate a safer world by reducing the number of nuclear weapons and by dramatically downsizing their military nuclear enterprises, but they can actually demonstrate that some of the excess materials of the Cold War—plutonium and uranium—can now be used to make energy for the people of the world. I emphasise the role that scientists can play because, as Academician Khariton pointed out to me, we all go back to the same roots. The conference on nuclear physics being held here concurrently with the workshop demonstrates how the scientific problems are the same, regardless of the native language of the scientists. It was said some time ago that "religion preaches the brotherhood of man, but science practices it." I believe that. And I am delighted to be here with all of you to consider these important scientific and social issues

# RADIOECOLOGICAL ISSUES IN THE RESEARCH EFFORTS OF THE RFNC-VNIIEF

R.I. ILKAYEV
*Russian Federal Nuclear Centre (RFNC)*
*All-Russian Scientific Institute of Experimental Physics (VNIIEF)*
*Arzamas-16, Sarov, Russia*

## 1. Introduction

Recent years have shown a significant increase in economic activities in the western areas of the Kazakhstan Republic where intensive efforts are underway for developing oil and gas condensate fields. This development has led to the question of whether it is possible to use industrially the underground cavities that have resulted from the peaceful nuclear explosions that occurred previously in that area.

This question has focused a special interest on the Lira site located about 140 km to the east from Uralsk where in 1983-1984 six explosions produced underground storage cavities. The explosions, which were conducted in strong rock salt at a depth of about 900 m, used a special nuclear charge of 15 kt of equivalent TNT, with a minimum of residual radioactivity [1]. The resulting five underground cavities of about $10^5 m^3$ each (the sixth cavity was filled with underground water after the shot), when used to store oil or gas, undoubtedly could offer a definite economic benefit. Therefore these cavities are of great interest to researchers [2].

The basic issue resulting from the actual use of these cavities would be the radioecological purity of the territory as well as that of the products (oil, gas, etc) that would be used after storage in these cavities. Such an assessment would require extensive radioecological research on the technology sites to obtain confident data indicating the potential presence of long-lived radionuclide products of nuclear explosions in this area and inside the underground cavities. This information would provide the basis for concluding whether or not these cavities could be used.

The significance of these issues is reflected in the materials of the first scientific and practical conference on the integrated studies and monitoring project of the Lira sites held in 1998 at the Nuclear Physics Institute of the National Nuclear Centre of the Kazakhstan Republic [2]. In addition, a great contribution to the solution of these problems can be made by the radioecological studies started in the territory of the former Azgir Test Site by the specialists from the Kazakhstan Republic, Russia, and other countries [6].

*S.S. Hecker et al. (eds.), Nuclear Physical Methods in Radioecological Investigations of Nuclear Test Sites*, 11-15.
© 2000 *Kluwer Academic Publishers. Printed in the Netherlands.*

# 2. Role of the RFNC-VNIIEF in the Radioecological Monitoring of the Lira Sites

## 2.1. MONITORING PROCEDURE

The procedure for the radioecological monitoring carried out by the Russian Federal Nuclear Centre (RFNC) of the All-Russian Scientific Institute of Experimental Physics (VNIIEF) at the previous nuclear explosion sites is well known and includes the following steps:

1. Visual inspection of the explosion site as well as the evaluation of the actual state of the neighbouring territory and equipment installed at the borehole mouth.
2. *In-situ* measurements using highly sensitive dosimetry devices and portable *in-situ* gamma spectrometers to define the level and the composition of the potential contamination by radioactive products. The measurements include the dose rate of gamma radiation, the contamination of the ground surface and the equipment with beta-emitting radionuclides. The results of gamma spectrometry are used to identify individual radionuclides.
3. Sampling of the ground surface over the entire technology area and of the equipment and buildings near the borehole mouth. The subsequent radiometry and radiochemistry analysis of the samples defines the level and structure of the contamination.
4. Analysis of the data obtained, comparison with the existing norms, and conclusions about the radioecological situation in the given area.

However, it is much more difficult to inspect the underground facility itself. These efforts require special boring equipment and diagnostic devices intended for operation under high pressures, high temperatures, and in challenging environments. Special geophysical studies are required to evaluate the actual state of the borehole and the cavity. It is quite possible that about 15 years after the explosion, partial collapse of the cavity dome chimney, and fissure zone took place.

## 2.2. MONITORING CAPABILITIES

The VNIIEF capabilities in performing the joint radioecological monitoring of the Lira sites result from the experience gained from similar activities carried out in the Russian territory. One example is the participation of the Institute in the research efforts relating to the investigation of the consequences of the radioactive fallout from the Chernobyl accident. As early as May 1986, the Institute obtained the first experimental results for the isotopic composition and concentration of radioactive materials dispersed along the radioactive cloud trace. Later the activities followed several additional lines.

A great number of measurements used the most sensitive methods to identify heavy elements (uranium, plutonium, americium, curium, etc.) in the chemically complex environmental objects of soil, plants, water, food, and biological compounds, etc. Special methods have been developed [7] that are based on the radiochemical separation of the analysed elements, their purification from the impurities, and their subsequent measurement, using state-of-the-art alpha spectrometers, gamma spectrometers, mass spectrometers, and other analytical equipment. The high sensitivity and accuracy of radionuclide identification in the samples are provided by the control of the complete separation with isotopic tracers. This process is used for identifying uranium, neptunium, plutonium, americium, and curium that were represented by the compounds of highly enriched isotopes obtained by the Institute on the electromagnetic separator S-2 [8]. For example, the radiochemical plutonium identification uses isotopically pure Pu-242 as a tracer. The concentrations of other isotopes are as follows:

$^{239}$Pu/$^{242}$Pu < $8 \cdot 10^{-6}$ Bq/Bq;
$^{240}$Pu/$^{242}$Pu = $1.6 \cdot 10^{-4}$ Bq/Bq;
238Pu/242Pu = $3 \cdot 10^{-4}$ Bq/Bq;
244Pu/242Pu < $10^{-9}$ Bq/Bq.

The identification limits of some radionuclides in the soil samples are much lower than the background (global) values owing to the presence of those isotopes in the environment. These isotopes, which resulted from human activity were as follows:

plutonium, americium, curium ~$4 \cdot 10^{-6}$ Bq/g;
stronium-90 ~0.02 Bq/g;
caesium-137 ~$10^{-4}$ Bq/g;
uranium ~$10^{-12}$ g/g scaled to the surface concentration: ~$10^{-5}$ Ci/km$^2$, ~$5 \cdot 10^{-2}$ Ci/km$^2$, ~$3 \cdot 10^{-4}$ Ci/km$^2$, ~0.1 g/km$^2$, respectively.

These research efforts conducted in co-operation with Ukrainian scientific institutions resulted in identifying concentration values in the surface layer of soil (at a depth of up to ~15 cm) in Ukrainian regions for such biologically significant radionuclides as $^{239}$Pu, $^{240}$Pu, $^{238}$Pu, $^{241}$Am, $^{90}$Sr, and $^{137}$Cs. These data provided the basis for the map of levels and distribution of radioactive contamination over the entire Ukrainian territory and separately over Kiev after the Chernobyl accident.

The VNIIEF also intensively investigates the gaseous form of this radioactivity. The use of radioactive tritium as a component of nuclear weapons required major efforts to develop appropriate technologies and safe manipulation measures. It was also necessary to create the corresponding hardware and methods to be used through all the phases of the technology cycle during the making and testing of nuclear weapons. In addition, relatively detailed studies were carried out on the structure of the gaseous products of underground nuclear explosions and on the behaviour of these products in the event of their potential leakage to the surface through the soil fissures and tamper systems. These studies also analysed the other characteristics important for understanding the processes occurring in the explosion cavity and in the environment. High-sensitivity experimental methods were developed to study the gaseous explosion products. These studies also used the isotopic tracers, that is the compounds of stable isotopes of krypton, xenon, and helium, in which one of the isotopes was enriched up to ~90%: $^{82}$Kr, $^{84}$Kr, $^{29}$Xe, $^{132}$Xe, and others.

Detailed studies were performed for the chemical composition of the gaseous phase of the explosion products. The computational codes and theoretical models were developed so that the evolution of the gaseous phase, the variation of the explosion products structure as they penetrate the rock fissures and release to the surface, could be tracked. The gas studies used a wide range of the state-of-the-art analytical capabilities, gas chromatography methods, mass spectrometry, liquid scintillators, etc. in combination with various gas purification methods.

A high sensitivity level was achieved for the developed recording methods. For example, for tritium ~$10^6$ at/sample for inert isotopes (Kr, Xe, Ne, He) ~$10^{10}$ at/sample.

A new and successful application for the efforts and capabilities of RFNC-VNIIEF is the implementation of radiation monitoring of the region where the atomic submarine "Komsomolets" sank in the Norwegian Sea. This programme was accomplished in 1992-1995 in co-operation with other Russian institutions [9].

As part of the above-mentioned programme, the RFNC-VNIIEF developed a hardware and methods programme for deep-sea radioecological monitoring of the sites of potential accidental sinking of carriers of nuclear power systems and nuclear weapons as well as for monitoring radioactive waste. The programme included the following equipment:

- Mobile version of the modified mass spectrometer MI-1201B for isotopic analysis of uranium and plutonium,
- Deep-sea semiconductor gamma spectrometer GAR-01 with a sensitivity of $10^{-11}$ Ci/l (caesium-137),
- Low-background semiconductor alpha spectrometer with a sensitivity of about $10^{-15}$ Ci/compound,
- Mobile version of the radiochemistry laboratory for fast measurements on board ship,
- General-purpose, in-depth probes for sea water and suspensions (particle size not greater than 1μm),
- Special sorbents for fissile materials,
- Additional devices.

The operation capabilities and technical potential of the programme were tested during a 1994-1995 expedition on board the research ship and carried out at the site of the sinking of atomic submarine "Komsomolets" in the Norwegian Sea. These technical efforts, carried out at extremely low depths, made it possible to detect, reliably identify, and define the concentrations of a number of natural and man-made radionuclides in sea water and sediments taken inside and near the submarine. The natural radionuclides of uranium, thorium, and the products of their decay: Ka-40, man-made materials—plutonium isotopes: $^{238}$Pu, $^{239}$Pu, $^{240}$Pu, as well as long-lived fission products: $^{137}$Cs and $^{90}$Sr were determined. The trace quantities of radioactive products, typical for the "Komsomolets" reactor: $^{134}$Cs, $^{137}$Cs, $^{54}$Mn, $^{90}$Sr, were detected. Normally, the measured concentrations of radionuclides corresponded to the background values of the world's oceans. These values in combination with their data made it possible to conclude that there is no significant impact of the submerged submarine on the environment in this region of the Norwegian Sea.

## 3. Conclusions

The outline of the monitoring programme discussed above shows that RFNC-VNIIEF has many years of experience in the organisation and implementation of radioecological studies in various media. These studies have been carried out at ground surface (radioactive contamination after the Chernobyl accident), under ground (in the case of underground nuclear explosions), and in the world's oceans (after the sinking of the "Komsomolets" submarine).

The Institute's state-of-the-art methods and hardware make it possible to compare the trace quantities of radionuclides, that are the products of nuclear explosion, with the technogenious radioactive materials existing in the environment naturally or as a result of technology. It seems reasonable to continue the Institute's international co-operation to resolve the problems associated with the industrial use of cavities previously formed by the underground nuclear explosions.

## 4. References


1. Soviet nuclear testing, vol.1. RFNC-VNIIEF, Sarov, 1997.
2. Integrated studies and monitoring of Lira sites, the 1st scientific and engineering conference, Almaaty, Nuclear Physics Institute, Republic of Kazakhstan, December 1-2, 1998.
3. Kedrovskii O.L. The use of underground nuclear explosions in industry and building.IAEA-PL-388/19, 20, 21. Peaceful Nuclear Explosions. Proceedings of a Panel Vienna, 2-6 march 1970, pp 17-21, 163-184, 377-404.
4. Hamburger R. United States of America, Statements on National programmemes. IAEA-PL-388/13. Peaceful Nuclear Explosions. Proceeding of a Panel Vienna. 2-6 March, 1970, pp.27-34.
5. Myasnikov K.V., Leonov E.A., Romadin N.N. The development of science and engineering base for the creation of underground storage facilities using underground nuclear explosions in rock salt. PNE-III: Proc. of a Panel (Vienna, 27 Nov- 1 Dec 1972) - Vienna: IAEA, 1974, IAEA-PL-388/19, pp 179-192.
6. V.B.Adamskii, Zh.I.Adymov, E.Z.Akhmatov et al. Peaceful nuclear explosion at the salt dome site Bolshoi Azgir, Preprint ¹ 10-98, Kazahstan Nuclear Physics Institute, Almaaty, 1998.
7. The guidelines. Evaluation of the concentrations of strontium, plutonium and americium radionuclides in soil samples (VNIIEF). M.: Gosgidromet SSSR, 1991.
8. S.P.Vesnovskii, B.N.Polynov. Nucl.Instrum.Meth., B70, p.9-11, 1992
9. Ocean studies and underground engineering activities at the submersion site of "Komsomolets" submarine. Edited by M.E.Vinogradov, A.M.Sagalevich, S.V.Khetagurov, M.,Nauka, 1996.

# TECHNIQUES USED AT THE NATIONAL NUCLEAR CENTRE'S INSTITUTE OF NUCLEAR PHYSICS FOR STUDYING SOIL PROBES AND SOIL FRAGMENTS AT THE SEMIPALATINSK NUCLEAR TEST SITE

K.K. KADYRZHANOV, V.P. SOLODUKHIN, I.V. KAZACHEVSKIY, S. KHAZHEKBER, S.N. LUKASHENKO, G.N. CHUMIKOV, M.F. VERESTCHAK, A.ELISEEV, A.K. ZHETBAEV, L.M. NAZARENKO, S.P. PIVOVAROV, A.PLATOV, A.B. RUKHIN, T.A. SEREDAVINA, P.V. CHAKROV, E.M. YAKUSHEV
*Institute of Nuclear Physics,*
*National Nuclear Centre, Republic of Kazakhstan*

**Abstract**

This paper presents a description of the specifics of the complex techniques used at the Kazakhstan National Nuclear Centre Institute of Nuclear Physics (KNNC INP) to solve the different problems connected with the investigation and radioecological situation at the Semipalatinsk Nuclear Test Site (SNTS). The results of applying these techniques are presented.

## 1. Introduction

After the prohibition of nuclear weapon tests in 1991, the task of investigating the radiological situation in the Semipalatinsk Nuclear Testing Site (SNTS) was becoming extremely urgent. Consequently, at the Kazakhstan National Nuclear Centre Institute of Nuclear Physics (KNNC INP) some activities were undertaken to create an appropriate instrumental, scientific, and technical base. It was quite natural that the main aim was the creation of a modern analytical laboratory capable of providing researchers with valuable scientific information on testing site contamination by artificial radionuclides and toxic elements. At the initial step of these studies, it became clear that the testing site contamination is rather peculiar, so nonstandard scientific and technical solutions as well as nontraditional radiological techniques are required. In particular, it has become clear that the existing standard radiochemical techniques developed and approved by governmental bodies in view of the determination of the contents of $^{239+240}$Pu (further $^{239}$Pu), $^{90}$Sr, $^{241}$Am originating from global precipitation are inadequate here. However, the absence of information concerning the exact nature of these radionuclides makes it difficult to accurately assess the situation at the testing site. The related studies have



*S.S. Hecker et al. (eds.), Nuclear Physical Methods in Radioecological Investigations of Nuclear Test Sites,* 17-42.
© 2000 *Kluwer Academic Publishers. Printed in the Netherlands.*

made it possible to establish that in many highly contaminated spots at SNTS, radionuclides in soils are characterised by rather strong chemical bonds. This fact strongly hinders quantitative chemical extraction by the acid leaching procedures used most often within the framework of approved radiochemical techniques. This peculiarity has made it necessary to study the forms of radionuclide occurrence in soils taken from various SNTS areas.

In addition, this information is required when studying migration properties of radionuclides as well as choosing the method for rehabilitating contaminated areas. The solution of so considerable and complicated a problem has required, in turn, attraction of corresponding techniques for soil sample fractionating as well as thorough studies of fractions and isolated fragments (grains) of soil samples. With all these circumstances taken in mind, a complex instrumental, scientific, and technical base capable of providing a variety of solutions related to the investigation of the radiological situation at the SNTS has been created and developed at the KNNC INP with the financial support of various international institutions (the International Atomic Energy Agency (IAEA), the International Science and Technology Centre (ISTC), etc.). This paper presents a brief description of the techniques used, their peculiarities, as well as their capacities regarding the solution of the SNTS problems. Radionuclide assay, being the most important, is discussed most thoroughly.

## 2. Experimental Methods

At the KNNC INP, the base facilities used for radiological studies are the following: the Y-150 isochronous cyclotron, the UKP-2 cascade charge-exchange heavy ion accelerator, the WWR-K nuclear reactor (KNNC IAE), a complete set of spectrometry instruments, the "Bruker" ESP-300E EPR spectrometer, the CM2201 MGR spectrometer, the AMRAY 1200 electron microscope, and the JY-70P atom emission spectrometer.

The technique's complete set incorporates the following:

- Radionuclide (instrumental, radiochemical) analysis;
- Element assay, including: activation analysis (AA), X-ray fluorescent (XRF) analysis ;
- Mass spectrometry (MS);
- Electron Paramagnetic Resonance Spectrometry (EPRS);
- Nuclear Gamma Resonance Spectrometry Spectrometry (NGRS);
- Electron microscopy (EM);
- Proton Induced X-Ray Emission (PIXE) with microprobe.

### 2.1. RADIONUCLIDE ANALYSIS

This technique incorporates the $\alpha/\beta/\gamma$ spectrometry of both artificial and natural radionuclides. Spectrometry is implemented by the "Canberra" 7401 $\alpha$-spectrometer having two chambers, the "Progress-BG" $\beta$-spectrometer (developed at VNIIFTRI,

Moscow), four γ-spectrometers provided with semiconductor coaxial/planar-type detectors, including the "Ortec" GEM-20180, "Canberra" GX-1520, "Ortec" GLP-25325, "Canberra"GLP-1010R.

*2.1.1. Instrumental Analysis*

This technique is realised by following common procedures of γ-spectrometry analysis. The essence of the technique consists of the following steps: the activity of the air-dried prepared sample (crushed or divided into granulometric fractions) is measured at the γ-spectrometer. Two weights having masses of 100 and 200 g are selected for complete sample and, as a rule, the whole fractions is extracted from an initial sample having a mass of 1 kg. The weight of 100 g is measured by the γ-spectrometer provided with a planar detector to determine $^{241}$Am (59.5 keV), $^{152}$Eu (121.8 keV), $^{154}$Eu (123.1 keV) and, in some cases, $^{137}$Cs (661.4 keV). The weight of 200 g is measured by the spectrometer that is provided with a coaxial detector for determination of $^{137}$Cs, $^{152,154}$Eu, $^{60}$Co (1173.2; 1332.5 keV), as well as natural radionuclides (U, Th, $^{40}$K). This methodical approach is associated with the following features: strong absorption of low-energy gamma-emission originating from $^{241}$Am and $^{152,154}$Eu, as well as considerable nonuniformity of Am distribution over the SNTS soil. High concentration in some grains makes difficult quantitative determination in the samples having masses larger than 100 g and heights greater than 10 to 15 mm, because the absorption function has a rather complicated form, considerably unlike the $1/g^2$ dependence.

For nuclides emitting more energetic gamma quanta ($^{137}$Cs, $^{60}$Co, etc.), measurements of samples having a mass of 200 g and the height of 20 to 40 mm do not introduce significant inaccuracies in determining its concentrations. Spectrometer information processing is performed at a rate of PC-486/100 by means of the code developed at the INP. The code implements peak search and identification, the determination of peak area with the statistical error and the error associated with peak interference taken into consideration, the calculation of the corrections to geometrical sizes of detector/sample/ various absorbers, and, finally, the calculation of the absolute activities on the basis of the quantum yields for the basic energy transitions. The results are presented as the values of the specific activity for each nuclide (Bq/kg). The technique makes it possible to determine the concentrations for the radionuclides listed above with the determination limit comprising 0.1 Bq/kg (for $^{241}$Am, 3 Bq/kg) and an error of 5 to 7%. The interlab experiment carried out in 1998 demonstrated good convergence of the results with the data of other participating laboratories (Kazakhstan National Nuclear Centre, Institute of Radiation Safety and Ecology (KNNC IRSE, the KNNC IAE, Radium Institute V.G. Khlopin – in Saint Petersburg).

Modern spectrometer instruments allow the considerable widening of a circle of the problems solvable by various instrumental techniques. In particular, modern planar-type detectors possessing high-quality characteristics make it possible to implement plutonium instrumental determination. The analytical characteristics of the "Ortec" GLP-25325 gamma-spectrometer with planar detector have been studied. The spectrometer makes it possible to measure soft γ/X-ray emission beginning with 3 keV. As is known, radioactive decay of long-lived plutonium isotopes (238, 239, 240, 241,

242) is not accompanied by any somewhat intense gamma emission; the emission intensity comprises, as a rule, the hundredth and thousandth fractions of per cent. The K-series X-ray emission is also of low intensity; more intense (3 to 4%) lines of the L-series have the energies 13.6 and 17.2 keV. This fact hinders strong application of the instrumental technique of plutonium determination in environmental objects. However, the complexity of plutonium radiochemical extraction makes it desirable to support any attempts to develop the instrumental techniques of determination. In this case, the determination for real objects plays a role in the main parameter. The major problems are the correct consideration of soft emission absorption in a sample as well as the contribution of the L-series lines originating from $^{241}$Am, which is always present in real samples. Measurement of the plutonium preparations after their radiochemical extraction have shown that the spectrometer efficiency (including the emission quantum yield) for the plutonium L-lines 13.6 and 17.2 keV comprises 0.1%. Then the plutonium determination limit (to be more exact, its modulus), estimated for the measurement time of 10 hours, comprises 0.5 Bq. Strong absorption of emissions by sample material does not allow the measuring of weights heavier than several grams. Successive measurements have shown that the optimum weight value for the detector in use comprises 5 g; then the value of the plutonium determination limit equals 1000 Bq/kg. At present, similar measurements are carried out by the spectrometer with the "Canberra" GL-1010R planar-type detector having better resolving power than the GLP detector and a higher (by a factor of ~2) sensitivity surface (respectively, 230 and 320 eV for 5.9 keV $^{55}$Fe). In Figure 1, for comparison, the spectra for one SNTS soil sample, taken with the detectors GLP and GL, are shown. It is quite evident the GLP detector provides a lower limit for plutonium instrumental determination.

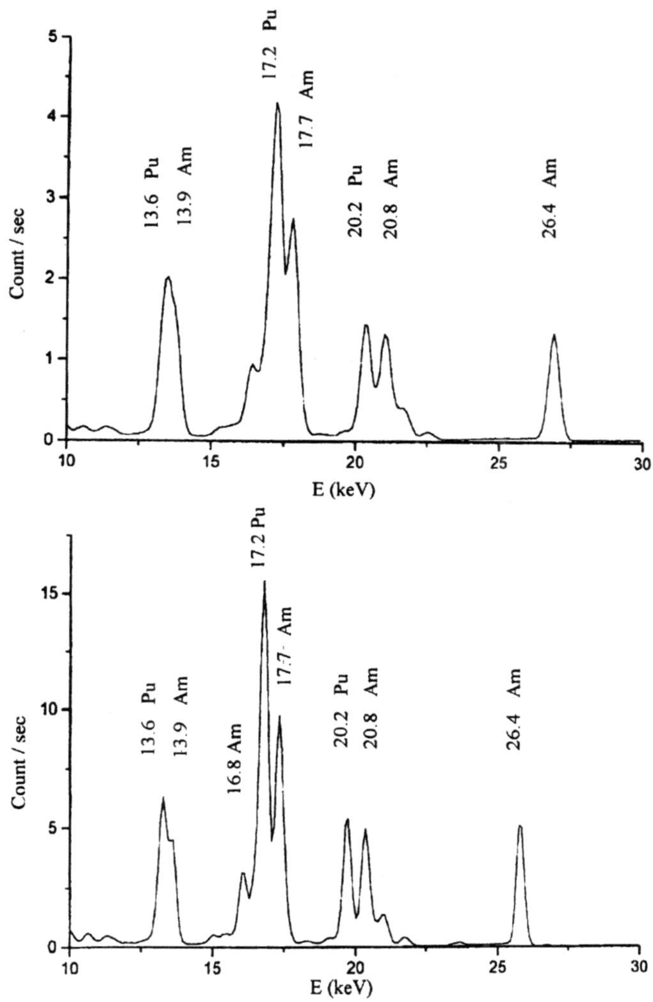


*Figure 1.* Gamma-spectrum of the soil sample (m=16 g, h=2 cm), measured with the detectors GLP25325 "ORTEC" (top) and GL1010 "CANBERRA" (bottom).

Although the studies performed have shown that the instrumental technique of plutonium determination is not capable of replacing radiochemical extraction in all cases, it can be rather effective for solving problems of a certain class. In particular, as a result of the investigation of the samples taken in the "Degelen" area, plutonium has been determined by means of the instrumental technique in samples with the highest activity. Also the instrumental technique is used for the implementation of screening "hot particles" in granulometric fractions.

*2.1.2. Complex Technique for Radiochemical Extraction of Pu, Am, and $^{90}Sr$*

Development of the complex radiochemical technique for determination of Pu, Am, and $^{90}$Sr in a single sample was dictated by the need to analyse single soil fractions having mass values insufficient for separate radionuclide determination. Besides, the overwhelming majority of the techniques for determination of the $^{90}$Sr concentration follow this scheme: Sr extraction – Y separation – $^{90}$Y collection – extraction of radiochemical pure $^{90}$Y (without $^{90}$Sr) – measurement of the $^{90}$Y β-emission. Determination of radiochemical yield in these techniques is carried out on the basis of the yields of Y/Sr carriers, which can be a source of significant errors. The process of establishing radioactive equilibrium between $^{90}$Sr and $^{90}$Y requires two weeks of settling, i.e., a very long analysis period. In most of the cases, known certified techniques for radioisotope assay, as a rule, are intended for determination of only one isotope and, sometimes, do not take into account complex forms of isotope occurrence in studied objects. Then application of this or that chemical stripping technique is required.

Consequently, when developing the technique for Pu/Sr/Am determination in environmental objects, we tried to provide the possibility for determining radionuclides in the objects of different contamination levels, both at a level of global precipitation and at the levels of contamination generated by nuclear tests, i.e., in differing forms of occurrence, sorption, and migration in natural systems. Besides, the developed technique is to be universal and capable of determining radioisotopes in such differing objects as water, plants, soil, biological materials, following a unique analytical scheme.

This problem has been solved by developing the technique for determination of such radionuclides as $^{239}$Pu, $^{90}$Sr, $^{241}$Am in environmental objects: all three radionuclides are determined by a single sample weight following the same analytical scheme (Fig. 2). Radiochemical extraction consists of transferring the radionuclides (Pu, Y, Sr, Am) from a sample to nitrogen acid solution. Then in the latter, radionuclide separation is carried out by the extraction chromatography technique at columns with successive receipt of a spectrometric source for every radionuclide. In the course of technique development, regimes for each step have been thoroughly studied. As tracers, such isotope labels as $^{88}$Y, $^{85}$Sr, etc. are used (The isotopes listed above are produced at the cyclotron A-150 by means of the techniques developed at the KNNC INP). Final determination of the radionuclide concentrations is performed on the basis of the α-spectrometry ($^{239}$Pu, $^{241}$Am) and β-spectrometry ($^{90}$Y – for determination of $^{90}$Sr). Complete analysis (determination of all three radionuclides) covers four days. Reproducibility of the results of the Pu, Am, and $^{90}$Sr determination has been verified with real soil samples taken at various SNTS areas. Discrepancy between the results for a sample does not exceed 10%. Besides, correctness of the determination of $^{90}$Sr has been tested with the standard sample IAEA-373 (Grass): the certified value 1312 Bq/kg has been determined as (1324±50) Bq/kg. The correctness of determining $^{241}$Am has been tested with the standard sample IAEA -135, the discrepancy if the certified and determined values has comprised 16%. Similar determination limit discrepancies have been obtained when compared with the results of γ-spectrometric determination of $^{241}$Am. The results of plutonium determination have been compared with those obtained at the Radium Institute (Saint Petersburg). The differences between the obtained values

– (179±16) Bq/kg and (151±14) Bq/kg – is quite satisfactory. The determination limits in the SNTS soil for these radionuclides are as follows: 10 Bq/kg for $^{90}Sr$, 0.5 Bq/kg for $^{239}Pu$, 1 Bq/kg for $^{241}Am$. The technique is protected by the author's certificate [1] and was certified by the enterprise "Gosstandard RK." One can find more details on this technique in Ref. [2].

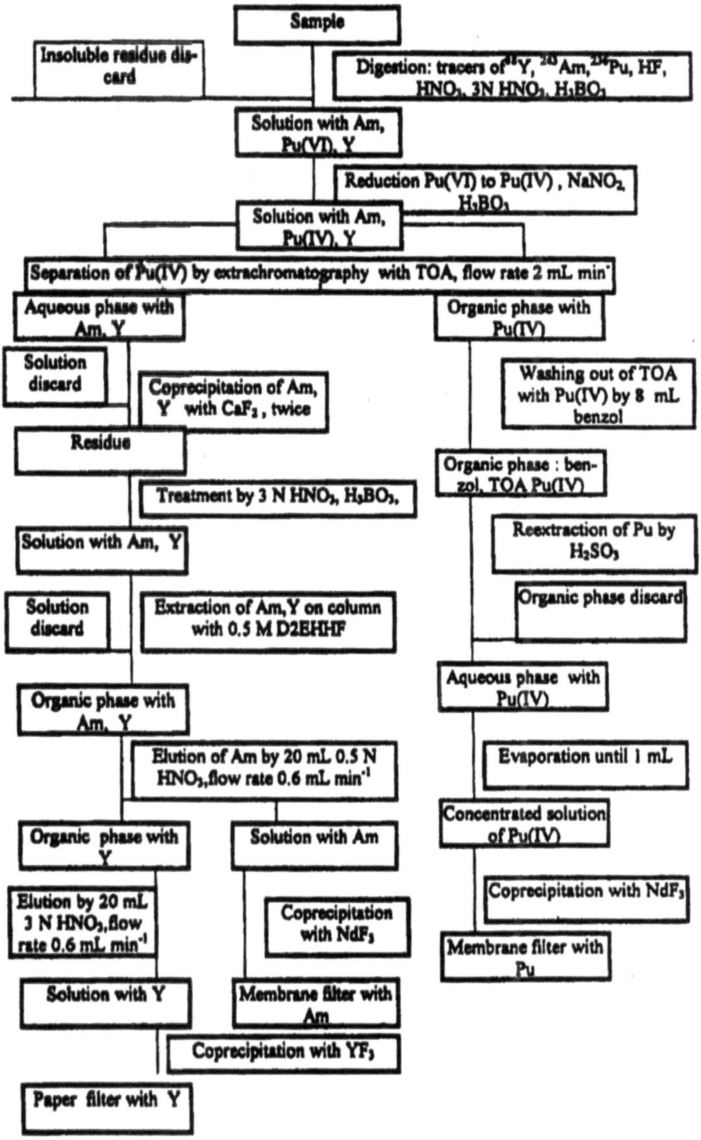


*Figure 2.* Scheme of the radiochemical separation of Pu, $^{90}Y$, Am.

The developed techniques of radionuclide analysis are used when solving all practical problems of radiological studies.

### *2.1.3. Element Analysis*

Determination of the element assay in the SNTS soil/soil fraction samples is implemented by activation analysis (AA) and X-ray fluorescent (XRF) analysis. These methods have been developed at the INP for more than 35 years and are assimilated. The techniques used in radiological and ecological studies have no principal differences in comparison to those developed earlier [3,4]. Consequently, in this section only a brief description of the analytical scheme and the basic devices are given. In view of activation analysis (AA) (determination of microelement content), the samples having the mass values of 0.2 to 0.5 g are irradiated in the experimental channel of the WWR-K reactor during three hours with the neutron flux comprising $2 \cdot 10^{13}$ $n \cdot cm^{-2} \cdot s^{-1}$. Measurements of the induced activity in samples are implemented in several steps with coaxial/planar-type $\gamma$-spectrometers. This analytical scheme makes it possible to determine more than 25 elements. A variety of methodical and instrumental capabilities is available with a view to widening the list[a], if required. The XRF analysis[b] is carried out at the facility incorporating the ampoule source of $^{109}Cd$ and the spectrometer with a semiconductor detector. Before measurement, samples are placed in standard dishes with bottoms made of thin polyethylene film. Spectrum processing is carried out by means of the computer code developed at the INP, which takes into account the sample absorbing features. This technique makes it possible to implement determination of 10 to 15 elements with the determination limit comprising 10 ppm. Some assays are carried out at the facility with a powerful X-ray tube (W=1 kW) with the application of various re-emitters (Fe, Cu, Zn, Mo, Cd). This trick makes it possible to reduce the determination limits for some elements to 0.1 ppm [4].

In addition to the techniques listed above, the GY-70P atom-emission spectrometer with induction-coupled plasma is available at the INP and also can be used for element assay [5].

Results of element assay are used to determine the content of various toxic contaminators[c] and the correlation between the soil/soil fraction element and radionuclide composition.

### *2.1.4. Nuclear Gamma-Resonance (NGR) Spectroscopy (Mössbauer Effect)*

In the initial stage of radiological studies at the INP, an interesting effect was revealed when separate fractions of the soil samples taken from the most contaminated areas were studied. Practically all of the Am and up to 80% of Pu were found to be concentrated in a so-called magnet fraction, extracted by means of soil treatment by a permanent magnet of 0.3 T. Consequently, the investigation of the iron forms/states in SNTS soil and its relations to products of nuclear tests was staged. A hypothesis

---

[a] Preliminary concentrating; radiochemical separation; various versions of activation by thermal/epithermal neutrons; analysis by short-lived radionuclides with application of rabbits, etc.

[b] In view of macroscopic composition determination.

[c] In particular, heavy metals.

regarding potential existence of unusual compounds of fission fragments and products of nuclear reactions with iron was also formed. In addition to chemical compounds, the formation of the systems related to fusion, sputtering, and mutual implantation at the conditions of super-high temperatures and pressures that accompany nuclear explosion also result. Nuclear gamma-resonance analysis seems to be the most promising technique for identification of the iron hypergenous compounds. It can be used to determine iron-containing minerals and their relative content in separate fractions as well as to determine the proportions between the iron forms with different valences.

In view of the Mössbauer studies, the samples of soil and their fractions are subject to grinding to 20 μm[d], mixed with paraffin and pressed to pellets with a thickness of 60 mg/cm$^2$.

Measurements performed with the Mössbauer (SM) 2201 spectrometer, equipped with cryostats and furnaces, make it possible to perform measurements with a temperature range from 4 to 800 K. Temperature impacts on crystalline lattice lead to variations in the hyperfine structure of the resonance nucleus energy levels, reflecting the chemical state, crystalline environment, electron structure, and dynamics of the Mössbauer atom [6,7].

Spectrometer control as well as the spectrometric information processing and mapping are implemented by means of the personal computer KOMPAN AT-Turbo. The Mössbauer source of $^{57}$Co having the activity ~1 mCi is used as the source of gamma quanta. The experimental spectrum processing is carried out by means of the computer code UNIVERSAL-UNIV. In this code, the sum of square deviations of the calculated Mössbauer parameter values from the experimental ones is minimised.

### *2.1.5. Electron Paramagnetic Resonance (EPR) Spectroscopy*

The EPR technique possesses high sensitivity against iron compounds. Also it provides the opportunity for separation of the ferromagnetic and paramagnetic fractions as well as for assessment of the symmetry characterising the surrounding paramagnetic complex. Then it follows that use of the EPR-spectroscopy data can facilitate understanding of the studied radionuclide nature and occurrence forms. Besides, the technique can be used for "radiation load reconstruction." The results of EPR studies of the soil taken near the nuclear explosion epicentres support this idea. Practically, for all these samples, the presence of a typical signal is inherent within the g-factor value 2, which is, rather, caused by radiation-induced centres in the solid matrix of a mineral. There is one more circumstance that requires special attention regarding relatively radiogenic paramagnetic centres. As these centres are metastable structures and possess high chemical activity, they can, when entering the human body, stimulate various chemical and biochemical reactions, including cancerogenous and mutagenic ones. All facts listed above strongly speak in favour of applying the EPT technique radiological studies.

At the INP, complex analytical equipment incorporating modern spectrometers[e], personal computers, and software for EPR spectrum processing has been assimilated.

---

[d] Larger grains induce nonresonant absorption.

All EPR studies of soils/soil fractions are implemented by means of these spectrometers [8-10]. Spectrum processing is carried out directly in the spectrometer with the code ESP 300E designed for the OS-9 operational system, as well as by the code *WINNER*, firm "Bruker." At the personal computer the files containing spectra and parameters are transferred to by cable and transformed for operation in the medium of Windows 95.

In the course of methodical studies, the influence of magnetic heterogeneity in the samples of SNTS soil and its separate fractions on resonance absorption have been investigated. The most simple and convenient way of control for the extent of heterogeneity in composition of the studied sample is its rotation in a resonator. In the course of rotation, parameters of the signal originating from isotropic sample must not vary. It has been found that turning the ampoules with the samples in question leads to a variation in the g-factors of the EPR lines, including narrow lines seemingly associated with paramagnetic centres (PMC) of radiation origin. On the basis of this fact, one can guess that the paramagnetism source is localised in separate fine particles of the SNTS soil. To verify this hypothesis, successive separation of the sample by means of a strong Co-Sm magnet has been used. In Figure 3 the absorption spectrum of the primary sample of the soil fraction is labelled as Fr 47 (the farthest in the figure, with a weight of 120 mg), the spectrum obtained after the first separation (the middle one in the figure, weight of 120 mg), the spectrum after repeated separation and seeding fine magnetic fraction off (the closest fraction, weight of 29 mg) are shown. After one separation, the intensity of the secondary wide line of the ferromagnetic resonance (FMR) reduced by a factor of 2.8 of the integral amplitude, by a factor of 3.1, of the area of the absorption line, by a factor of 3.4, the FMR line width has been changed from 1687 to 1037 gauss. The resolving power has not increased, and line a proves it. In the closest spectrum, considerable narrowing the wide line (to 540 gauss) and reducing its amplitude are observed. The major peculiarity of the spectrum is the considerable increase of the resolving power within the range of strong local magnetic fields, i.e., poor fields for recording. As a result, at least, three narrow lines (a, b and c) are markedly observed, and the width of the line a, having the g-factor ~4.3, which is present in many spectra, is reduced by a factor of 4.7 (to 15.9 gauss). The last fact is very important for the interpretation of its origin.

[e] The spectrometers *ESP 300E, firm "Bruker"* and *EPR10-mini.*

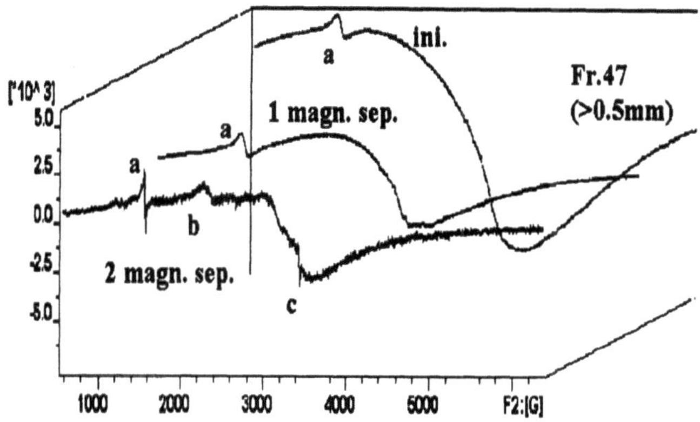


*Figure 3.* Spectra of the soil fraction sample Fr 47, taken before magnetic separation (initial), and after single and double magnetic separation.

On the basis of the methodical experiments performed, the following conclusion has been made: Before thorough investigation of the PMC signals in the then studied sample, preliminary sample grinding and magnetic separation are required. Precautionary measures are to be taken in order not to distort the sample paramagnetism pattern.

*2.1.6. Mass Spectrometry*

Among radionuclide contaminators, transuranium elements (Pu and Am) occupy a special place, because they possess high radioactive toxicity. The application of traditional radionuclide assay techniques, when studying isolated soil fragments, fails in the cases associated with restriction by the absolute sensitivity (the determination limit $\sim n \cdot 10^{-11}$ g). Besides, these techniques are incapable of providing determination of the isotope composition. As for the techniques based on mass spectrometry possessing high sensitivity with respect to the elements in question (the determination limit $\sim n \cdot 10^{-14}$ g), this problem is quite solvable; moreover, isotopes can be determined separately.

Mass spectrometry of heavy radionuclides requires the application of complicated and rather expensive facilities. Standard instruments are not fitted to these measurements. Special developments are needed. For example, in Reference 11, in view of the provision of high isotopic sensitivity within the uranium/transuranium element mass range, the mass spectrometric tandem consisting of two sector magnets of the total ion-optic tract length is more than 6.5 m and the ion beam energy keV is proposed. At the KNNC INP, a simpler mass spectrometric installation has been created, and high isotopic sensitivity is achieved owing to new prism ion optics deflecting/focusing elements that provide a better quality of beam focusing [12, 13]. In Figure 4, the ion-optics scheme for the developed mass spectrometer facility is depicted.

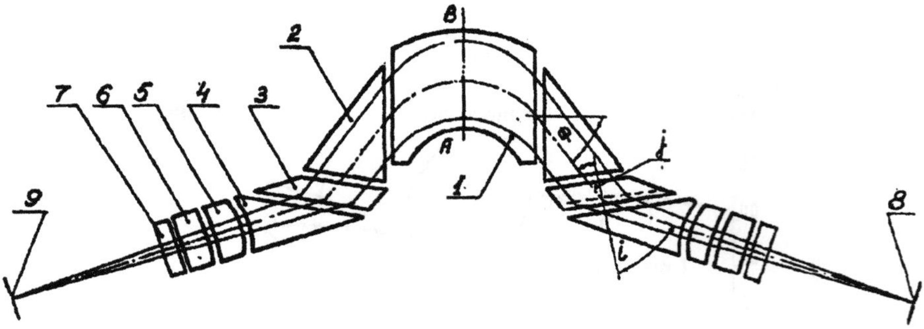


*Figure 4*. The ion-optics scheme of the mass spectrometer facility.

The plane AB divides the ion-optical system into two identical parts. The magnetic prism represents the electromagnet with flat-parallel pole pieces. Number 1 is equipped with two magnetic screens; number 2 provides confinement of the scattered field as well as its two-dimensionality.

The ion beam focusing in energy is achieved by the deviating cylindrical lenses formed by electrodes 2 to 4 in two identical three-electrode telescopic systems. The electrodes (playing also a role of magnetic screens) are grounded. Ion deflection is implemented by the electric two-dimensional fields disposed within the gaps between electrodes 2 and 3, and 3 and 4. The fields also focus the ion beams in the vertical direction. Telescope capability of the entire system is achieved by applying the potentials to electrodes that provide coincidence of the linear foci of the cylindrical lenses (the back focus of the first lens and the front focus of the second one). Electrodes 4 to 7 form the collimating lenses and the focusing lenses. In the focus of the first lens, exit slot 8 of the ion source is located; whereas in the focus of the second lens, the entrance slot 9 of the receiver is disposed.

The appropriate calculations and experiments have been performed in order to determine the main parameters of the ion-optical scheme, including: the ion-beam angle of incidence to the magnetic prism $\phi=51°$, the ion beam angle of refraction in the first telescopic system $j=31°15'$, the angle of incidence in the same system $i=64°42'$ and the focal length of the front (collimating) and back (focusing) lenses F=561.6 mm. With the parameters listed above, the linear dispersion in the mass for the developed mass spectrometer, comprising $D_m=F\cdot tg\phi\cdot cos\ j/cos\ i=1157$ mm, is close to the value of the dispersion for the tandem mass spectrometer [4], equal to 1200 mm. It should be noted that this large value of the dispersion in the prism instrument is achieved with the length of the ion-optical tract comprising 1.6 m, being almost a factor of 3 less than that in the sector device. With the parameter values given above, the beam in the course of its transport from the source to the receiver undergoes a five-fold refocusing, which is similar to the periodic focusing in the charged particle accelerator. As a result, the beam is confined near the mid-plane, and in the course of beam transportation from the source to the receiver, the ion current losses are kept to a minimum.

The application of prism ion optics has made it possible to create a unique mass spectrometer facility possessing high isotopic sensitivity in the mass range corresponding to transuranium elements: the determination limit for $^{239}Pu$ comprises $2 \cdot 10^{-14}$ g.

*2.1.7. Electron Microscopy*

In the course of radionuclide composition studies of the soil samples taken from various SNTS areas, the existence of considerable irregularities in radionuclide distribution over the sample volume has been established. Further studies have revealed the availability of the so-called "hot particles"—the finest particles having sizes not greater than $n \cdot 100$ μm and high radionuclide concentration. Information on the structure of these particles and the forms of radionuclide occurrence in them is quite useful for determination of forms of radionuclide occurrence in the SNTS soils. Consequently, in view of the investigation of the hot grain microstructure and element composition, the electron microscopy techniques incorporating local element assay and microscopic proton-induced X-ray emission (PIXE) have been used.

Modern techniques of electron microscopy make it possible to obtain surface images with the resolution power to 10 nm and extremely high sharpness of the picture. That is, it is possible to carefully study the morphology of particles and the various kinds of heterogeneity in its structure (pores, cracks, inclusions of various phases) within a wide range of the magnification values, ranging from several units to several tens of thousands. The extremely tiny size of the electron probe offers the chance to record the spectra of the characteristic X-ray emission from the area of 1×1 μm and, thus, to determine quite accurately the element composition for rather small segments of the studied objects.

In view of "hot particle" microscopy, a special technique of sample preparation has been developed. The technique incorporates the huge particles crushing (grain size is greater than 0.1 μm) and fixing on a metallic substrate of 5×5 mm by means of thermoplastic glue and the thin electricity conducting layer carried onto the surface by vacuum deposition. A sample prepared like this is placed in the microscope chamber. Investigation is commenced with the determination of the variation in grain morphology (when observing with low magnification (about ×100)). If considerable morphology variation is observed, the examined particles are divided into several groups, and further careful studies at high magnification (up to $\times 5 \cdot 10^{4}$) are performed with 3 to 5 particles per group. Identification of the depicted structure is accompanied by the record of the X-ray spectra generated by local spots (from 100×100 μm to 1×1 μm), and special attention is paid to typical morphology details. When soil fractions are studied, they are divided into the "magnetic" and "nonmagnetic" fractions; in turn, the nonmagnetic fractions are divided by their mass into "light" and "heavy" [14]. The studies are implemented by the Philips SEM 515 scanning microscopes provided with the energy-dispersion analyser LINK AN 10000/85 and the JCXA 733 with the wave X-ray spectrometer. At present the Amray 1200 electron microscope is being used.

When the "hot particle" structure and composition are studied, three regimes of image used most frequently are:

- Secondary electron contrast image regime that provides information on grain morphology and surface relief;
- Reflected electron contrast image regime that provides information on differences in the average atomic number over various areas of a sample;
- Contrast image regime for the characteristic X-ray uranium emission, which provides information on uranium distribution over the sample surface.

The semiquantitative element assay of chosen surface areas having sizes from 1×1 to 100×100 μm is carried out by means of the wave X-ray spectrometer either at the focused probe regime or at the scanning one.

*2.1.8. Proton-Induced X-Ray Emission Technique (PIXE)*

The PIXE technique is based on the capability of the accelerated protons to excite the inner electron shells inside an atom. On analysing the characteristic X-ray emission (CXE), one can determine its element composition. The PIXE technique possesses certain advantages compared to the similar techniques in which CXE excitation is induced either by X-rays or by electrons. The absence of scattered/bremsstrahlung X-ray emission makes it possible to increase considerably the assay sensitivity. In the microscopic PIXE version, the strictly focused beam (~1÷50 μm) is used, making possible the study of the spatial distribution of elements over the surface of a studied object.

At the INP the microscopic PIXE technique is realised on the basis of the tandem accelerator UKP-2-1. The accelerator possesses the following operational parameters:

- accelerating voltage — up to 1 MW
- voltage instability — less than 100 V
- energy spread — less than 200 eV
- proton current in the beam — up to 60 μA
- proton energy — up to 2 MeV

The microbeam line consists of the following elements: the object slots serving as diaphragms for the ion-beam transverse size; the secondary slots disposed at some distance from the object slots and intended for extraction of the beam's central part with low angular divergence; the scanner that implements transverse scanning over the sample surface; the lens strictly focusing the beam to a small spot and the target chamber. All units are assembled at the accelerator proton tract exit on the massive support coaxial to the beam.

The diameter of the beam obtained within the target plane comprises 15 μm with the current in the sample equal to ~2 nA for the scanned area comprising 500×500 μm. The target chamber used for analysis is provided with eight changeable sample holders, the optic microscope and the Si(Li)-detector being arranged at an angle of 30° relative to the beam [15].

Test experiments have been carried put. The diagrams of distribution of such elements as Ca, Mg, Ti, Si, Al, Mn, Fe, Cr and U over the surface of a soil grain have been constructed.

## 3. Results and Discussion

All the techniques described above are used successfully at the KNNC INP for investigation of radiological situation and for contaminator characterisation at the SNTS.

According to the character of the studies performed, all these techniques can be divided into two categories: macroanalytical and microanalytical. All techniques of the radionuclide/element analysis (AA RPA) as well as NGR/EPR-spectroscopy refer to macroanalytical methods. EM (electron microscopy), the micro-PIXE analysis and MS (mass spectroscopy) refer to microanalytical ones. However, it should be said the classification like this is rather conventional, as when solving real problems, macro-techniques in microanalytical studies and vice versa are often used.

### 3.1. MACROANALYTICAL STUDIES

The techniques of radionuclide analysis are applied for solving practically all problems. The main results obtained with attraction of these techniques are presented in this Proceedings volume under the title: "Characterisation of Radionuclide Contamination of the Territory of the Semipalatinsk Nuclear Testing Site" by K. K. Kadyrzhanov, *et al.* and in References 16 through 19.

Application of the techniques of element assay (AA and XRF) to the investigation of soil samples taken from various SNTS areas has made it possible to reveal a number of peculiarities. By means of the XRF, macrocomposition of the soil samples taken along the southeastern tail of radioactive precipitation, from isolated areas of the Balapan area and in the vicinity of the town of Kurchatov has been studied. The concentration intervals uncovered for such elements as K, Ca, Ti, Cr, Mn, Fe, Ni, Cu, Zr, Rb, Sr, Y, Zr, Nb, Pb, Th, and U bear witness to the relative homogeneity in soil composition for the areas studied. However, in some adjacent areas and even in various layers of a sample taken from one place, contrasting features in the contents of some elements are observed. Table 1 shows the results of determination of the element composition in the soil generalised samples (20 cm) and surface (5 cm) samples taken from the southeastern track (No. 1 and No. 6), the southwestern track (No. 5), and in the "Balapan" area near "Chagan" lake (No. 9). The values of the concentration for many elements in these samples differ by a factor of 1.5 to 2. The presented data prove also that in three areas (Nos. 1, 6, 9), the element concentrations in surface layers are considerably less than those in the appropriate generalised samples. One of the potential reasons for this phenomenon is element migration to the earth depth under the precipitation impact. The exception is area No. 5, where the element contents in the surface layer are substantially higher than those in the appropriate values in the averaged samples. This feature is associated, rather, with the high content of cinder on the surface of the area where the elements, practically, are not washed out.

TABLE 1. The element contents (in %) in the soil generalised samples ($C_g$) and the surface samples ($C_s$) taken from various areas of the SNTS.

| Element | Plot No. | | | | | | | | | | | |
|---|---|---|---|---|---|---|---|---|---|---|---|---|
| | 1 | | | 5 | | | 6 | | | 9 | | |
| | $C_g$ | $C_s$ | $C_g/C_s$ | $C_g$ | $C_s$ | $C_g/C_s$ | $C_g$ | $C_s$ | $C_g/C_s$ | $C_g$ | $C_s$ | $C_g/C_s$ |
| Ti | 1.483 | 0.967 | **1.53** | 1.584 | 2.467 | **0.64** | 2.138 | 1.419 | **1.51** | 2.234 | 1.457 | **1.53** |
| *Cr* | 0.311 | 0.181 | **1.72** | 0.233 | 0.388 | **0.60** | 0.436 | 0.255 | **1.71** | 0.369 | 0.270 | **1.37** |
| *Mn* | 0.274 | 0.150 | **1.83** | 0.312 | 0.522 | **0.60** | 0.320 | 0.204 | **1.57** | 0.344 | 0.214 | **1.61** |
| *Fe* | 4.303 | 2.924 | **1.47** | 6.603 | 6.822 | **0.97** | 4.823 | 3.066 | **1.57** | 5.981 | 4.262 | **1.40** |
| *Ni* | 0.035 | 0.035 | **1.00** | 0.056 | 0.064 | **0.88** | 0.061 | 0.049 | **1.24** | 0.065 | 0.036 | **1.81** |
| *Cu* | 0.050 | 0.031 | **1.61** | 0.048 | 0.068 | **0.71** | 0.073 | 0.040 | **1.83** | 0.080 | 0.048 | **1.67** |
| *Zn* | 0.034 | 0.025 | **1.36** | 0.032 | 0.042 | **0.76** | 0.042 | 0.028 | **1.50** | 0.049 | 0.030 | **1.63** |
| *Pb* | 0.013 | 0.010 | **1.30** | 0.016 | 0.017 | **0.94** | 0.023 | 0.012 | **1.92** | 0.022 | 0.010 | **2.20** |
| *Sr* | 0.053 | 0.039 | **1.36** | 0.048 | 0.055 | **0.87** | 0.063 | 0.036 | **1.75** | 0.095 | 0.054 | **1.76** |
| *Zr* | 0.034 | 0.025 | **1.36** | 0.039 | 0.039 | **1.00** | 0.041 | 0.026 | **1.58** | 0.042 | 0.030 | **1.40** |

On the basis of the data on the element/radionuclide composition of the soil samples taken from the southwestern tail, analysis of potential correlation between them has been performed. The obtained results speak in favour of the assumption of a relationship between the data on $^{241}$Am and $^{60}$Co, on one hand, and on Ti, on the other hand. This finding reveals that regularity can be explained by the formation of stable compounds incorporating radionuclides and the constructional material of products in the course of testing. In the Degelen area in the neighbourhood of the A-1 tunnel, the enhanced contents of such elements as Mn, Y, Nb and Mo have been found as well as an associative relationship between the element composition in the soil samples and the contents of $^{239}$Pu and $^{241}$Am in them: the direct relation—for Mn, Zn, Y and the reverse relation—for Fe and Zr. This phenomenon can be related to the peculiar composition of the water flowing from the adit and its effect on the soil composition along the water flow bed. Additional studies are required in order to study the phenomenon more carefully.

By means of the AA, microelement composition of the soil samples taken along the southeastern tail has been studied. The obtained results presented in Table 2 demonstrate that at a distance of 30 km from the epicentre, the contents of all elements are enhanced. One potential explanation involves the contamination resulting from regional specific technogenic activities.

The application of the NGR spectroscopy to the investigation of the soil/soil fraction samples taken from various SNTS areas has made it possible to identify several forms of iron occurrence in the samples studied. It has been found that in all fractions, iron is in the two/three-valence state. The two-valence iron presents in the form of sulfide, carbonate and silicates. The three-valence iron crystallises in the form of hematite, magnetite, and hötite, mainly, as small fractions, appearing on the fine disperse state of magnet-ordered iron in soils. Owing to application of the low-temperature Mössbauer studies of the soil samples taken near the explosion epicentre in the "Opytnoe Pole", hötite in the magnet-ordered state, i.e., having the grain size greater than 200 μm, has been revealed. The effect found supports the assumption that the soil thermal treatment leading to baking the fine dispersed particles of hötite into much larger aggregates took place as a result of the impacts accompanying nuclear explosions. This feature, in the case of confirming its nature, can be used for identification of the places where nuclear tests were performed.

TABLE 2. Distribution of element contents in soil at the southeastern track of SNTS

| Distance | Element contents (ppm) | | | | | | | | | | | | |
|---|---|---|---|---|---|---|---|---|---|---|---|---|---|
| *km* | **Fe, %** | ***Co*** | ***Ni*** | ***Zn*** | ***As*** | ***Sb*** | ***Nd*** | ***Sm*** | ***Eu*** | ***Tb*** | ***Tm*** | ***Yb*** | ***Lu*** |
| 0.7 | 4.60 | 18.0 | 32.30 | 113.3 | 8.0 | 3.0 | 40.7 | 5.1 | 1.15 | 1.03 | 0.5 | 2.0 | 0.74 |
| 6 | 5.26 | 28.0 | 51.40 | 115.3 | 22.0 | 3.1 | 30.3 | 6.7 | 1.38 | 1.34 | 0.6 | 3.0 | 0.93 |
| 8 | 5.37 | 20.0 | 23.90 | 113.9 | 17.0 | 2.9 | 47.4 | 5.2 | 1.21 | 1.16 | 0.7 | 2.0 | 0.91 |
| 10 | 4.94 | 17.0 | 35.90 | 91.2 | 4.0 | 1.7 | 34.0 | 4.7 | 1.20 | 0.74 | 0.5 | 1.0 | 0.59 |
| 12 | 6.61 | 22.0 | 29.90 | 105.5 | 4.0 | 2.6 | 17.0 | 4.9 | 1.26 | 1.04 | 0.6 | 2.0 | 0.79 |
| 14 | 6.00 | 19.0 | 19.10 | 92.6 | 13.0 | 1.6 | 20.0 | 4.3 | 1.12 | 0.96 | 0.3 | 1.0 | 0.65 |
| 16 | 1.71 | 7.0 | 13.20 | 36.6 | 3.0 | 0.8 | 17.8 | 4.7 | 0.90 | 0.90 | 0.5 | 2.0 | 0.72 |
| 18 | 4.42 | 18.0 | 43.10 | 113.3 | 3.0 | 1.7 | 38.5 | 6.2 | 1.30 | 1.07 | 0.9 | 2.0 | 1.13 |
| 24 | 3.45 | 14.0 | 33.50 | 60.4 | 14.0 | 1.8 | 40.0 | 4.0 | 0.96 | 1.07 | 0.5 | 2.0 | 0.83 |
| 30 | ***8.73*** | ***70.0*** | ***63.40*** | ***122.4*** | ***232.0*** | ***16.2*** | ***51.1*** | ***7.8*** | ***2.11*** | ***1.79*** | ***1.0*** | ***3.0*** | ***1.47*** |
| 34 | 4.12 | 15.0 | 33.50 | 82.7 | 10.0 | 1.7 | 21.5 | 4.8 | 1.22 | 1.06 | 0.7 | 2.0 | 0.90 |
| 40 | 4.11 | 13.0 | 27.50 | 71.9 | 7.0 | 2.4 | 25.2 | 5.3 | 1.28 | 0.96 | 0.6 | 2.0 | 1.02 |
| 50 | 6.08 | 23.0 | 37.10 | 81.7 | 23.0 | 2.8 | 45.1 | 6.4 | 1.58 | 1.26 | 0.5 | 3.0 | 1.16 |
| 58 | 7.43 | 38.0 | 34.70 | 115.3 | 20.0 | 3.9 | 20.0 | 6.4 | 1.55 | 0.90 | 0.3 | 2.0 | 1.05 |
| 68 | 5.06 | 27.0 | 44.30 | 97.7 | 21.0 | 2.8 | 30.3 | 6.8 | 1.45 | 1.45 | 0.6 | 2.0 | 1.09 |
| 90 | 4.61 | 16.0 | 21.50 | 63.1 | 14.0 | 1.8 | 31.1 | 5.2 | 1.15 | 0.96 | 0.6 | 2.0 | 1.00 |
| 112 | 4.21 | 15.0 | 22.70 | 79.0 | 4.0 | 1.1 | 13.3 | 4.3 | 0.99 | 0.72 | 0.4 | 1.0 | 0.68 |

In the course of the EPR studies, a variety of the spectra generated by samples of soil/soil fractions taken from various SNTS areas has been obtained.

The typical electron resonance spectrum of the SNTS soil is depicted in Figure 5. Here one can observe a wide signal of the ferromagnetic resonance caused by the ferromagnetic material embedments. The signal shape and intensity vary from sample to sample, thereby pointing to the presence of magnetic minerals, highly ferromagnetic magnetite $Fe_3O_4$ having in its central part the g-factor of about 9.0, weakly ferromagnetic hematite $Fe_2O_3$ having the g-factor of about 6.0, and, mainly, antiferromagnetic limonite FeOOH $nH_2O$ with the g-factor of about 3.0 in the samples studied. Also the signal near the value of the g-factor of ~4.3, caused by the $Fe^{3+}$ paramagnetic ion in the orthorhombic crystalline field is observed prominently. The most probable model of this ion is $FeO_2^{2-}$. Besides, in some samples, the characteristic signal of $Mn^{++}$ presenting in soil, as a rule, in a very small quantity (less than 1 ppm), manifests itself. It is a natural impurity that does not require special investigation, as do other EPR signals observed in the spectrum.

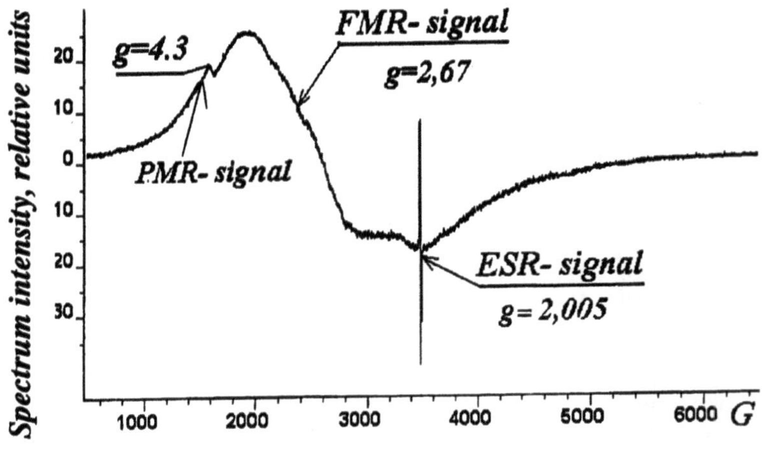


*Figure 5.* Typical ESR spectrum of SNTS soils.

A narrow signal in the region of g=2.0 is of the most interest. The radiation origin of the signal has been proved by special experiments with additional artificial irradiation. These experiments have proved the complete identity of the signals recorded in the SNTS soils in the region of the "Opytnoe Pole" or along the radioactive precipitation tail and in the signals observed in similar soils but far from the "Opytnoe Pole," and subject to special artificial irradiation.

Investigations of the dependence of the signal shape on the value of the microwave power as well as the studies in the Q range have shown that the signal possesses a rather complicated nature and represents the superposition of the signals generated by several paramagnetic centres. The spectrum record at a very low power level has made it possible to isolate a specific signal caused, seemingly, by the "$E'_1$"–centre in $SiO_2$; however, this conclusion needs additional support.

Investigation of the signal distribution over mineralogy fractions has shown that it is accumulated in different fractions with different efficiency (Fig. 6). This conclusion is supported in the experiments (in addition to the studies of the particles of different colours) on the investigation of various natural minerals and rocks in the SNTS.

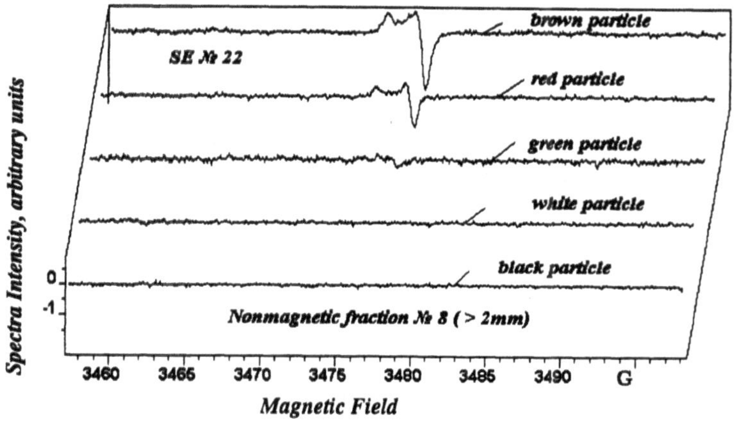


*Figure 6.* Stack of the EPR spectra generated by soil grains of different mineralogical composition (different colours).

## 3.2. MICROANALYTICAL STUDIES

The studies of the microstructure and microcomposition of the SNTS soil isolated particles form this category of investigation. The main techniques are: electron microscopy, microscopic PIXE, and mass spectrometry. Radionuclide analysis, activation analysis, and EPR spectroscopy also compose the complete set of the applied techniques.

The investigation of the isotopic composition of the magnetic fractions extracted from the soil samples taken from various SNTS areas has been carried out by the mass spectrometry technique. In some particles of these fractions, the isotopes of Ra, U, Pu, Am have been revealed. In Figure 7, for illustration, the mass spectrum of a single grain of the magnetic fraction extracted from the soil sample taken in the "Opytnoe Pole" is presented. The lines corresponding to the isotopes of $^{224,226,228}$RA, $^{234,235,238}$U, $^{239,240}$Pu, and $^{241}$Am are clearly seen. It has been established that uranium is enriched with light isotopes: $^{235}U/^{238}U$ =0.162 (natural ratio ~0.007); $^{234}U/^{238}U$ =$1.4 \cdot 10^{-3}$ ($5.77 \cdot 10^{-5}$). In the course of investigation of the isotopic composition in single particles belonging to the same sample, considerable heterogeneity in the distribution of the Pu/Am radionuclides over the sample of magnetic fraction is revealed, being, probably, associated with the peculiarities of the physical and chemical processes that accompany nuclear tests.

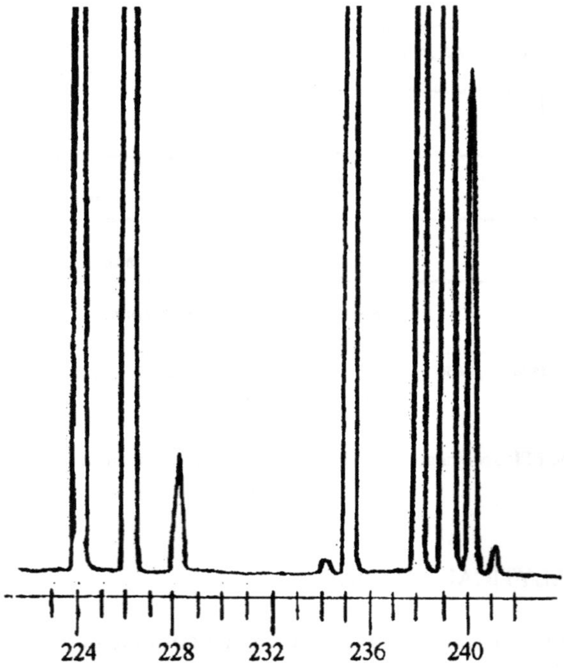


*Figure 7.* The mass spectrum of the magnetic fraction for one of the SNTS soil samples ("Experimental Field").

With the electron microscopy technique, the investigation of the microstructure and microcomposition of various particles[f] of the SNTS soils has been implemented. It has been established that the ferromagnetic properties of magnetic particles are caused by the presence of inclusions of the ferromagnetic phase with the high content of iron (more than 90 weight % relatively to $Fe_2O_3$). The nonmagnetic matrix of these gains is formed by minerals based on silicon (50 weight % relative to $SiO_2$) with impurities of Al, Mg, Na, K, Ca, Fe. The magnetic inclusion size varies within a wide range of values—from several tenths of a millimetre to several microns. Some part of its tiny magnetic particles (the fraction size from 0.05 to 0.063 mm) entirely consists of the magnetic phase and contains a large number of radionuclides per normalised weight. "Non-magnetic" particles, on average, have more heterogeneous structure; their composition, as a whole, is similar to that of the nonmagnetic phase of "magnetic particles." An insignificant fraction of nonmagnetic particles differs by the high content of Zr (more than 50 weight % relative to $ZrO_2$) with impurities of Si, Al, Pr and, seemingly, represents zircon.

"Melted" particles (Figs. 8a, b, c, d, e, f, g, h), by their structures, contain a large quantity of spherical pores (e) of various sizes, having smooth surface. Most consist of Si (about 70 weight % relative to $SiO_2$), Al, K, Fe, Ca, Na, Mg (g). On the surface,

[f] "melted," "magnetic," "non-magnetic," "hot particles," etc.

silicon "balls" and circular bulges of the same composition and of the size up to 200 μm are observed (e). In addition, the circular "spots" that are not observed against the major mass in relief but have a different composition with enhanced content of either Fe or Ca (f, h) are revealed.

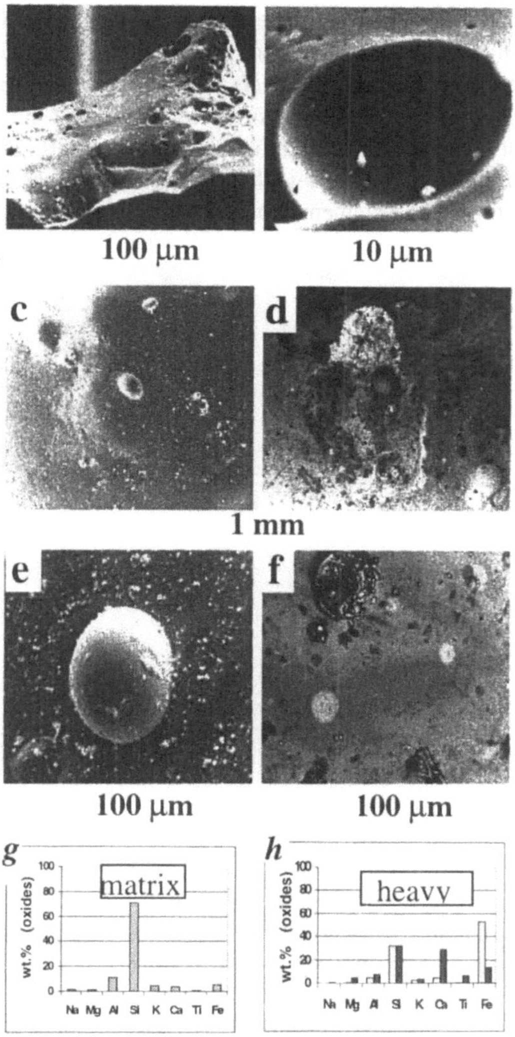


*Figures 8 (a-h).* Large glassy-melted particle from SNTS soil, SEM images, and composition diagrams.

The structure and composition of "hot particles" have been studied. A grain has a flat shape, its width and length comprising about 400 μm, the thickness about 100 μm. According to the data of γ spectrometry, the grain contains 370 Bq of $^{239}$Pu. One of the flat surfaces has been studied by the scanning microscope. Three regimes for image receipt have been used: SEI (secondary electron image)[g], BEI (backscattered electron image)[h] and the contrast regime of the characteristic X-ray uranium emission. Also semiquantitative element analysis of the chosen segments of the sample surface has been carried out. The general pattern of the grain under studies is shown in Figure 9 (a, b). The material matrix is formed by the elements lighter than sodium, specifically, by carbon (probably in graphite form) with an insignificant impurity of Ca, Si and Al. On the surface the inclusions having high average atomic numbers (i.e., heavy elements) are observed. The majority is of the size of up to 50 μm and the element composition typical for the most widespread natural minerals. The "heaviest" inclusions (by contrast under the BEI regime) have a size of less than 10 μm. Among them, the following types of inclusion, differing by their element composition, are present:

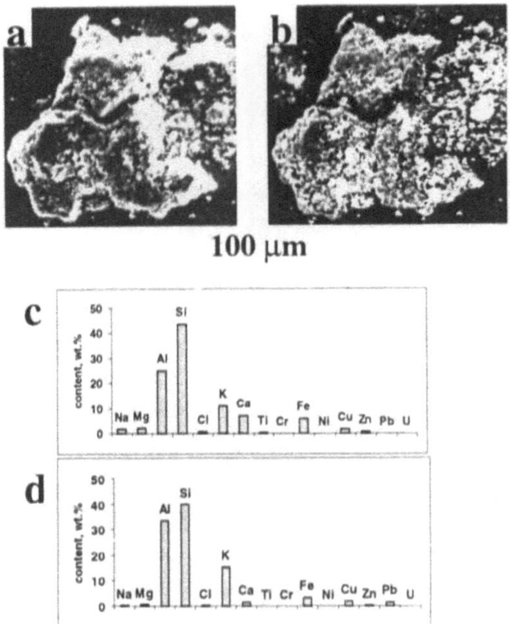


*Figures 9 (a-d).* Electron microscopic image of the hot grain and the diagrams representing its composition over various grain segments.

1) Inclusions containing a considerable amount of lead with impurities of lighter elements, without signs of uranium or plutonium;

---

[g] The secondary electron contrast image regime that provides information on grain morphology and releif of its surface.

[h] The backscattered electron contrast image regime that provides information on the differences between the average atomic numbers for various sample segments.

2) Inclusions containing 25 to 50 weight % of uranium, as well as Pb, Zn, Cu, Ni, Fe and a number of lighter elements;

3) Inclusions consisting mainly of uranium (about 90 weight%) with an insignificant impurity of lighter elements (Figures 10 (a-g)).

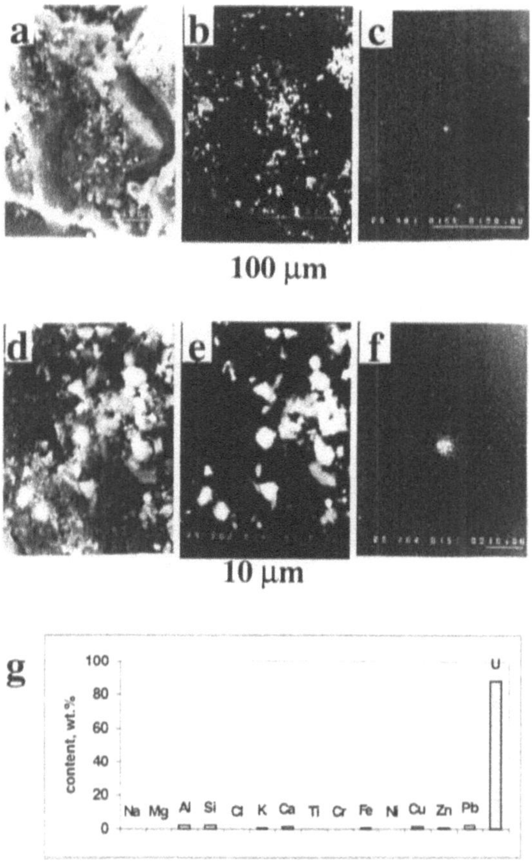


*Figures 10 (a-g).* Inclusion with a high content of uranium (in the centre) on the surface of the "hot" particle.

Qualitative analysis reveals the presence of plutonium in all inclusions containing uranium. Unfortunately, the instrument in use has not made it possible to measure its content quantitatively.

Special attention is to be paid to the high level of localisation of uranium and plutonium in the sample studied. These elements are concentrated in isolated inclusions of less than 10 μm rather than distributed more or less uniformly over the grain volume.

An attempt has been undertaken to study plutonium distribution over the hot grain surface by the microscopic PIXE technique. As a result, the PIXE spectrum has been measured in a point of the grain (Fig. 11).

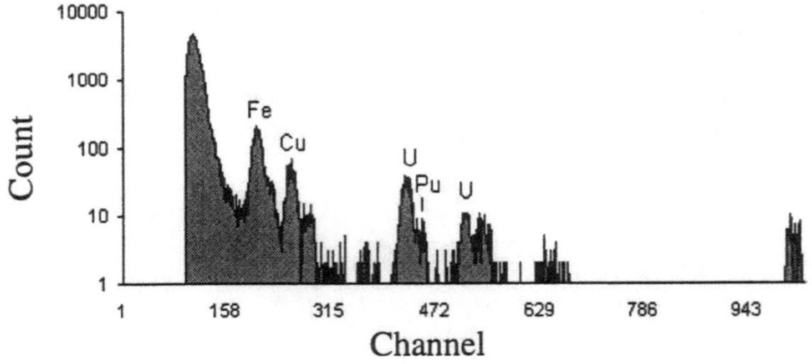


*Figure 11*. The PIXE spectrum of one of the surface areas in the hot grain.

In this spectrum prominent peaks of the uranium L-lines 14.3 keV (447 channels), 18.3 keV (546 channels) and 21.4 keV (661 channels) have been revealed. It is of interest that the plutonium peaks are present near the uranium ones. Then the cartogram of plutonium-uranium distribution over the grain surface has been constructed (Fig. 12, left side).

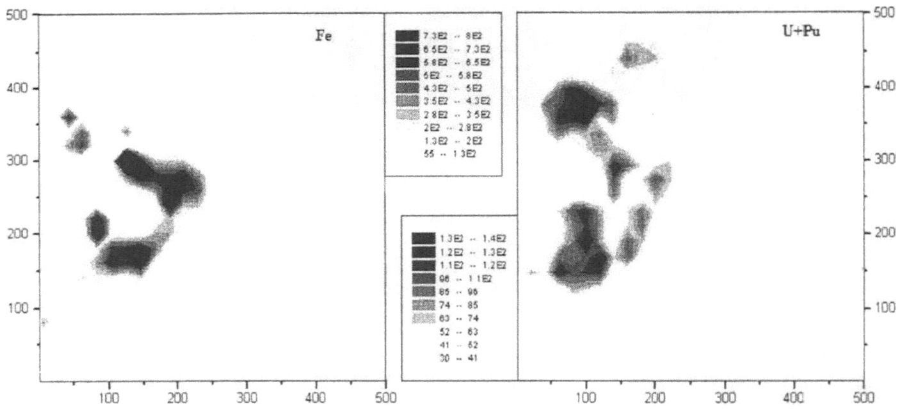


*Figure 12*. Distributions of uranium (left) and iron (right) in the SNTS sample. The beam diameter is ≈15 μm.

For comparison, the cartogram of iron distribution over the same surface has been constructed (Fig. 12, right side). The experiments performed point to the potential correlation between these three elements. Experiments to support this assumption are in progress.

## 4. Conclusions

Owing to the urgency of studying the negative effects of the nuclear tests at the SNTS on the environment and in view of the development of a scientifically justified approach to the elimination of these effects, a modern complex analytical method has been created at the KNNC INP. This system includes the following:

- An analytical laboratory capable of implementing the up-to-date determination of the radionuclide element composition in the environmental objects has been created and certified by Kazakhstan standardisation bodies.
- A technique for determination of such isotopes as $^{230}Pu$, $^{90}Sr$, $^{241}Am$ in soil, plants, and natural water has been developed. The technique has been protected by the patent for inventing the way for Pu/Am extraction from the SNTS soils and concentration.
- An instrumental base has been created and the technique has been developed for mass spectrometric determination of the isotopic composition of transuranium elements in the SNTS hot soil particles.
- In view of the studies on phase composition in the iron-containing fragments of soil taken from the radioactive precipitation tails, the NGR spectroscopy technique has been adapted to radiological studies.
- The EPR spectroscopy technique making it possible to reconstruct the dose loads at various nuclear testing site areas has been developed on the basis of the up-to-date instruments.
- At the UKP-2 accelerator, the PIXE microbeam technique for microprobing hot particles has been developed. Also, in view of the investigation of the microstructure/microcomposition of various soil particles, the technique of scanning electron microscopy has been developed.

These techniques are used successfully to study the levels of radioactive contamination and the state of radionuclide contaminators at various SNTS areas. The facilities under study are as follows: "Opytnoe Pole," where ground/air explosions were performed; the mountain range "Degelen," where underground explosions were performed in adits; the "Balapan" area, where underground explosions were performed in wells; the places where excavation explosions were carried out, such as "Chagan," "Telkem," "Murzhik," etc. All of the information obtained is recorded in a special radiological database. The database makes it possible to implement complex assessment of the radiological situation at SNTS, to imitate and forecast its evolution, as well as to optimise scientific and technical plans for future studies.

The complex radiological instrumental and methodical base created at the KNNC INP is quite universal and can be used for the investigation of radiological situations in other places where nuclear explosions were performed.

## 5. References

1. I.V. Kazachevskiy, A.K. Zhetbaev, V.P. Solodukhin, K.K. Kadyrzhanov, G.N. Chumikov, L.N. Smirin, S.N. Lukashenko, N.M. Berdinova "Way for extraction of concentrating plutonium and

americium out of the Semipalatinsk testing site soils" – in Russian. Kazakhstan Patent No.16084, 1995.

2. I.V. Kazachevskiy, S.N. Lukashenko, G.N. Chumikov, E.T. Chakrova, L.N. Smirin, V.P. Solodukhin, S. Khazhekber, N.M. Berdinova, L.A. Ryazanova, V.I. Bannyh, V.M. Muratova. Combined Radiochemical Procedure for Determination of Plutonium, Americium and Strontium-90 in the Samples from SNTS. Czech. J. Phys., 48 (1998), 69-84
3. V.V. Sokolskiy, V.P. Solodukhin, I.V. Kazachevskiy "Nuclear-physics analytical techniques for Kazakhstan science and industries" In Proceedings of the Int. Conf. "Nuclear and radiation Physics" Almaty, 1997, p.208-216
4. V.P. Solodukhin "Nuclear-physical methods of analysis in oil geology and industry of Kazakhstan". J. Radioanal. Nucl. Chem. 234 (1998) 101-105.
5. S.N. Lukashenko, N.M.Berdinova, G.N. Chumikov, V.P. Solodukhin , I.V. Kazachevskiy, L.A. Ryazanova. Microelement analysis of natural water by the atom emission spectroscopy technique with high-frequency inductively-connected plasma. – in Russian. Izv. AN Kaz. SSR, ser. Fiz.-mat., o. 4 (1989) 91-93
6. A.K. Zhetbaev, M.F. Verestchak, V.A. Donbaeva, T.D. Dosmaganbetov, Z.T. Serikbaeva, S. Khazhekber. Diagnostics of iron state in the SNTS soils by means of the Mössbauer effect technique. In Proc. of Int. conf. "Nuclear and Radiation Physics". Almaty, 1997, p.118
7. M.F. Verestchak, V.A. Donbaeva, A.K. Zhetbaev, S.P. Pivovarov, A.B. Rukhin, T.A. Seredavina, Z.T. Serikbaeva. Study of Soils of the SNTS by Means EPR and Mossbauer Spectroscopy. In Proc. of the 29th Int. Conf. Ampere-13th ISMAR, Berlin, 1998, p. 1065
8. S.P. Pivovarov, I.V. Kazachevskiy, T.A. Mosienko, V.P. Solodukhin, O.V. Stakhov. Anormalous Intensity of ESR-Signal from Glassed Soil Particles of SNTS Epicentre. 1st Asia-Pacific EPR-ESP Symposium, Hong Kong, 1997, p. 106.
9. S.P. Pivovarov, T.A. Seredavina, A.B. Rukhin. ESR of Environmental objects SNTS. – in Russian. In Proc. of Int. conf. On ESR Dosimetry, Obninsk, 1998, p. 114.
10. Kazachevskiy I.V., S.P. Pivovarov, Solodukhin V.P., S. Khazhekber, Chumikov G.N., Lukashenko S.N., A.B. Rukhin. Radionuclides Composition and Paramagnetism of Soils of the SNTS. 13th Radiochemical Conference, Jachymov 1998, 93.
11. Mc Cormick A. Thermal-Ionization Mass-Spectrometry for Small Sample Analysis of Uranium and Plutonium. Appl. Radiat. Isot., V.43, No. 1-2 (1992) 271-278.
12. V.M. Kelman, S.P. Karetskaya, L.V. Fedulina, E.M. Yakushev. Electron Optics Elements of the Prism Mass Spectrometers.- in Russian "Nauka", Almaty, 1979.
13. V.M. Kelman, L.M. Nazarenko, E.M. Yakushev. Theory of symmetric prism mass spectrometer – in Russian. Zh.T.F., v.42, No. 5 (1977), 963-968.
14. P.V. Chakrov, S.P. Vagin, S.V. Sedov, I.V. Kazachevskiy, S.N. Lukashenko. Analysis of morphology and element compositions of the soil particles out of the nuclear explosion tail. – in Russian. In Proc. of the 10-th All-Russian Symposium on Scanning Electron Microscopy and Analytical Methods of Investigation of Solids (•••-97)", p.100.
15. A.Arzumanov, A. Borisenko, I. Gorlachev, A.Eliseev, S.Lysukhin, A.Platov, A.Tuleushev, "Application of electrostatic tandem in Almaty for analysis of beryllium foils and radioactive particulate", Proc. of the 6 EPAC, Stockholm, June 1998, p.2433-2435.
16. I.V. Kazachevskiy, V.P. Solodukhin, S. Khazhekber, S.N. Lukashenko, G.N. Chumikov, L.N. Smirin "Investigation of radionuclide composition in the soils/soil fractions of the former nuclear weapon testing site". In Proc. of Int. Conf. "Nuclear and Radiation Physics". Almaty, 1997, p.96
17. I.V. Kazachevskiy, V.P. Solodukhin, S. Khazhekber, L.N. Smirin, G.N. Chumikov, S.N. Lukashenko, "Some aspects of determination of radionuclides at the former Semipalatinsk Nuclear Testing Site", J. of Radioanal. and Nucl. Chem., Vol.235, Nos 1-2 (1998) 145-149
18. L.N. Smirin, I.V. Kazachevskiy "Software for radiological studies". In Proc. of Int. Conf. "Nuclear and Radiation Physics". Almaty, 1997, p.102
19. V.V. Sokol'skiy, V.P. Solodukhin, I.V. Kazachevskiy. "The main direction of development of nuclear-physical methods of analysis for science and industry of Kazakhstan". "13-th Radiochemical Conference", Marianske Lazne-Jachymov, Chech Republic. 1998, p.205

# GENERAL ECOLOGICAL PROBLEMS OF NUCLEAR TEST SITES

# HEALTH AND ENVIRONMENTAL ISSUES AT U.S. NUCLEAR TEST SITES

L.R. ANSPAUGH
*Radiobiology Division*
*University of Utah*
*729 Arapeen Drive*
*Salt Lake City, UT 84108-1218*

**Abstract**

Of the 1030 nuclear tests the United States conducted from 1945–1992, 210 were in the atmosphere; the more significant sites were at the Bikini and Enewetak atolls in the Marshall Islands and at the Nevada Test Site (NTS). Ecological effects of these tests consist mainly of short- and long-term contamination of surface soil, contamination of groundwater, and surface disturbance. One test in the Marshall Islands resulted in the evacuation of 241 Marshallese, including 12 pregnant women, as well as 28 military weather observers. Twenty-three Japanese fisherman were also exposed. Many of these persons suffered acute radiation sickness, but no immediate deaths occurred. The collective dose to these highly exposed persons was about 200 person Sv; no estimate is available for lesser exposed persons. At the NTS no persons were documented to be highly exposed, but the collective dose to the population living in the nearby downwind area has been estimated to be 12 000 person Gy. Observed health effects from the tests have included acute radiation sickness in the heavily exposed group in the Marshall Islands: One late death was due to infection possibly caused by multiple blood transfusions, and there was an increased incidence of thyroid cancer. Health effects in the population around the NTS have been difficult to detect; there is a statistically significant increase in the incidence of acute leukaemia in a subgroup of a case-control study and there is a statistically significant association of thyroid neoplasms with dose, but not for thyroid nodules or for thyroid carcinomas.

## 1. Introduction

The United States (US) conducted its first test of a nuclear weapon prototype on 16 July 1945 at Alamogordo, New Mexico; a 21-kt explosion [1]. Even though this test was conducted under conditions of secrecy, its environmental effects did not go undetected owing to the contamination of straw in the Midwest that was used as packing material for photographic film [2]. From this beginning and until the last test conducted on



*S.S. Hecker et al. (eds.), Nuclear Physical Methods in Radioecological Investigations of Nuclear Test Sites*, 45-60.
© 2000 *Kluwer Academic Publishers. Printed in the Netherlands.*

23 September 1992, the United States conducted 1030 tests at a variety of locations as shown in Table 1: 815 tests were underground, 210 were in the atmosphere, and 5 were underwater. In addition, another 24 tests were conducted jointly by the United Kingdom and the US at the Nevada Test Site (NTS). Of the four tests in the Pacific Ocean two were underwater and two were at high altitudes; the three tests in the South Atlantic were high-altitude rocket tests. The 24 tests at Christmas Island and the 12 tests at Johnston Island were all conducted at high altitude. The tests at Carlsbad, Farmington, Grand Valley, and Rifle were Plowshare tests, i.e., they were mostly experiments in stimulating the production of natural gas. Underground tests to test the ability to detect nuclear weapons tests were conducted at Fallon (1), Hattiesburg (2), and one at Amchitka. Two underground "seismic calibration" tests were conducted at Central Nevada and Amchitka; the remaining test at Amchitka was an underground test of an antimissile warhead. The five tests at the Nellis Air Force Range, which adjoins the NTS, were non-nuclear Pu-dispersion tests.

TABLE 1. Nuclear tests conducted by the United States at various locations [1]

| Location | Number |
|---|---|
| Pacific Ocean | 4 |
| Johnston Island (Pacific) | 12 |
| Enewetak Atoll (Pacific) | 43 |
| Bikini Atoll (Pacific) | 23 |
| Christmas Island (Pacific) | 24 |
| South Atlantic | 3 |
| Nevada Test Site (underground) | 804 |
| Nevada Test Site (atmospheric) | 100 |
| Central Nevada | 1 |
| Amchitka, Alaska | 3 |
| Alamogordo, New Mexico | 1 |
| Carlsbad, New Mexico | 1 |
| Hattiesburg, Mississippi | 2 |
| Farmington, New Mexico | 1 |
| Grand Valley, Colorado | 1 |
| Rifle, Colorado | 1 |
| Fallon, Nevada | 1 |
| Nellis Air Force Base Range, Nevada | 5 |
| Total | 1030 |

The nuclear yields of the tests conducted by the United States have ranged from 0 to 15 000 kt (Shot Bravo conducted on 1 March 1954 at Bikini Atoll [1]); the larger tests (and many smaller ones) were conducted in the atmosphere until such tests ended in November 1962 [1]. The location and total yields of the US tests conducted in the atmosphere are shown in Table 2. Although the majority of tests were conducted at the NTS, in general, the atmospheric tests conducted there were of small yield (the largest was 74-kt Shot Hood conducted on 5 July 1957 [1]). All of the tests at Johnston and Christmas islands were at high altitude and thus resulted in relatively minor effects. Larger effects would be anticipated to have occurred at Bikini and Enewetak atolls where there were many high-yield tests conducted near the surface. Until the present time, the more severe environmental and health effects have been associated only with atmospheric tests, although at later times there may be effects associated with the movement of radionuclides with groundwater.

TABLE 2. Location and total yield of atmospheric nuclear explosions conducted by the United States. Data summarised from [1] to one significant digit

| Location | Yield (Mt) |
|---|---|
| Alamogordo, New Mexico | 0.02 |
| Nevada Test Site | 1 |
| Enewetak Atoll (Pacific) | 30 |
| Bikini Atoll (Pacific) | 80 |
| Pacific Ocean | 0.1 |
| Johnston Island (Pacific) | 20 |
| Christmas Island (Pacific) | 20 |
| South Atlantic | 0.004 |

All tests of larger yield injected some of their debris into the upper troposphere and even into the stratosphere; such injection resulted in so-called global fallout that was distributed throughout the hemisphere into which it was injected. Because it took some time (on the order of a few years) for debris injected into the stratosphere to return to Earth, most of the short-lived radionuclides had already decayed and the effects of global fallout were less than that of local fallout.

The purpose of this paper is to summarise and discuss the health and environmental effects of local fallout from tests of nuclear-weapons related devices at the US test sites. The effects of global fallout, which are less than those from local fallout, are not discussed. Also, while the effects of tests at the Pacific atolls are discussed, emphasis is placed here on the effects from tests at the NTS. This emphasis has been chosen for three reasons: one is that the effects of the tests at the Pacific atolls have recently been summarised elsewhere [3]; a second is that a much larger population is potentially at risk from the tests at the NTS; and, finally, the author's own research has focused on the NTS.

## 2. Ecological Issues

The ecological issues associated with nuclear weapons tests at US sites fall into three main categories: contamination by radionuclides of surface soil, contamination by radionuclides of groundwater, and land disturbance. Each issue has both short-term and long-term aspects. For tests at the Pacific atoll sites there are additional issues with contamination of sediments in the associated lagoon systems.

Immediately after the explosion of a nuclear device on or near the soil (or water) surface, the presence of many short-lived radionuclides produces a radioactive field that can be very intense. Fortunately, this radiation field decreases rapidly with time; as a first approximation, the decreasing exposure rate, $\dot{X}(t)$, can be described by a power function [4]:

$$\dot{X}(t) = \dot{X}(1)\, t^{-1.2} \ , \tag{1}$$

where $t$ is time in hours.

If the details of the device materials and construction are known, a more accurate representation with a sum of exponential functions can be made [5-7]. Over longer time periods only radionuclides with half lives of several years survive to create a residual radiation field for potential ingestion and/or inhalation of the radionuclides. Typically, the radionuclides of very long-term interest are $^{137}$Cs, $^{90}$Sr, and isotopes of Pu.

Most of the tests conducted by the United States were underground tests at the NTS. Fortunately, the distance from the surface at the NTS to groundwater is unusually far, although several tests were conducted in or near the groundwater. This creates a long-term potential for the movement of radionuclides with groundwater to sources of potable water.

The final impact of general interest is the disturbance of land, particularly land surface. This has been one of the greater impacts from the testing of nuclear devices.

## 2.1. PACIFIC ATOLLS

Inhabitants of the Bikini and Enewetak atolls were evacuated before the tests began (in 1946 on Bikini and in 1948 on Enewetak). The unfulfilled desire of these displaced people to make an unrestricted return to their ancestral homes represents perhaps the most significant negative impact of the United States nuclear testing programme [8, 9]. The reason that they have not been able to occupy their ancestral lands freely is the contamination of the land with residual radionuclides. The primary radionuclide of concern is $^{137}$Cs, which is avidly taken up by vegetation in this calcium rich, but potassium deficient, soil system [10, 11]. The second most important radionuclide is $^{90}$Sr. Each of the two radionuclides has a half life of about 30 years, so this problem will eventually be resolved by time, if not by other remedial efforts. The additional radionuclide of long-term concern is $^{239+240}$Pu; an extensive cleanup campaign[1] was conducted on Enewetak Atoll with the general goal of removing Pu down to levels of 40 pCi per gram [8]. In general Pu is the dominant radionuclide moving through the inhalation pathway, but this pathway contributes only a small fraction of the total dose; the more important pathways are ingestion and external gamma exposure [10].

The sediments in the lagoons at Bikini and Enewetak are also contaminated, and $^{137}$Cs, $^{60}$Co, and $^{207}$Bi have routinely been detected by gamma spectrometry in flesh samples of reef and pelagic fish; in general, however, the contribution of marine foods to the ingestion dose is small [12].

Land disturbance at the Bikini and Enewetak atolls has been severe, as many of the larger US tests were fired at these locations at or near the surface [1]. This resulted in

[1] The cost of the cleanup was $218 million, and six lives were lost in accidents.

the complete vapourisation of some of the islands, e.g., Elugelab Island on Enewetak Atoll [8]. In addition, there was extensive disturbance of the sediments in the lagoons. Another extensive land disturbance was associated with the plutonium cleanup on Enewetak and with the on-site storage of the debris in one of the craters that then received a concrete cap.

## 2.2. NEVADA TEST SITE

The NTS was created as the second continental test site (after Alamogordo, NM) by order of President Truman late in 1950. This action followed a decision to pursue the development of the hydrogen bomb and the perceived need for more rapid testing of experimental devices at a secure site (the Korean Police Action was then in progress) [13]. The first test was conducted on 27 January 1951 [1]. As indicated above, most of the US tests have been conducted at this site although the total yield of atmospheric tests was only about 1 Mt. The magnitude of the underground tests was never large, owing to concerns about seismic damage in the nearby metropolitan area of Las Vegas. A schematic map of the NTS is shown in Fig. 1 [14]. The site is about 100 km northwest of Las Vegas; it occupies a 3750 $km^2$ area that was carved from the Nellis Air Force Base Range Complex, which abuts the NTS on three sides. The Range Complex, which provides an additional buffer area, was also used as the site of five Pu-dispersal experiments. One hundred atmospheric tests were conducted at NTS with a total yield of 1 Mt. This resulted in the creation of 3 $10^{26}$ fission-product atoms, most of which decayed rapidly. Of the more significant radionuclides, 6 PBq of $^{137}Cs$ and 6 EBq of $^{131}I$ were released.

The NTS is of biological interest, as it straddles the boundaries of the Great Basin and Mojave deserts. Ecological impacts, while not as severe as those at the Pacific atolls, have resulted from the nuclear tests, as well as from the associated human intrusion and from natural events [15]. It has been estimated that approximately 25% of the NTS has been used for testing, including an extensive network of roads, support facilities, and areas fenced for the requirements of radiation control [15]. The most widespread disturbance has been caused by subsidence craters resulting from underground tests; of secondary significance is the vegetation denudation resulting from above-ground tests during the 1950s [15]. The most spectacular land disturbance is the Sedan Crater, a 390-m diameter, 100-m deep crater formed by a 104-kt Plowshare excavation experiment in July 1962 [1].

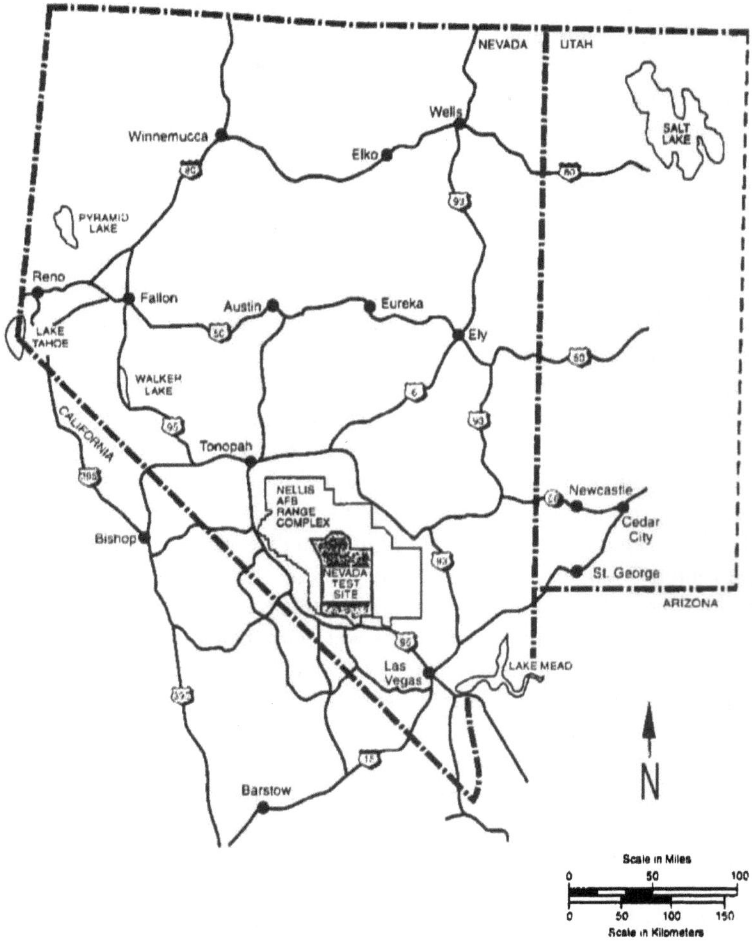


*Figure 1.* Location of the NTS and of the Nellis AFB Range complex. From [12].

Although the water table is deep (200 to 625 m) several tests have been conducted at depths near or below the water table [16, 17]. During underground tests a cavity forms (Fig. 2) that contains a "puddle" of refractory radionuclides; it is generally believed that the radionuclides contained in this puddle will not migrate, although the amount of positive evidence is small. Other radionuclides, particularly tritium, will move freely. From the analysis of pumping tests at nearby wells, the highest observed activity was indeed that of tritium, but elevated amounts of other soluble radionuclides including $^{90}Sr$ and $^{125}Sb$ were also observed [17]. On the basis of the observed measured concentrations in nearby wells and a model of radionuclide transport, it was concluded that it was very unlikely that significant concentrations of any radionuclide would reach the nearest source of useful groundwater [17].

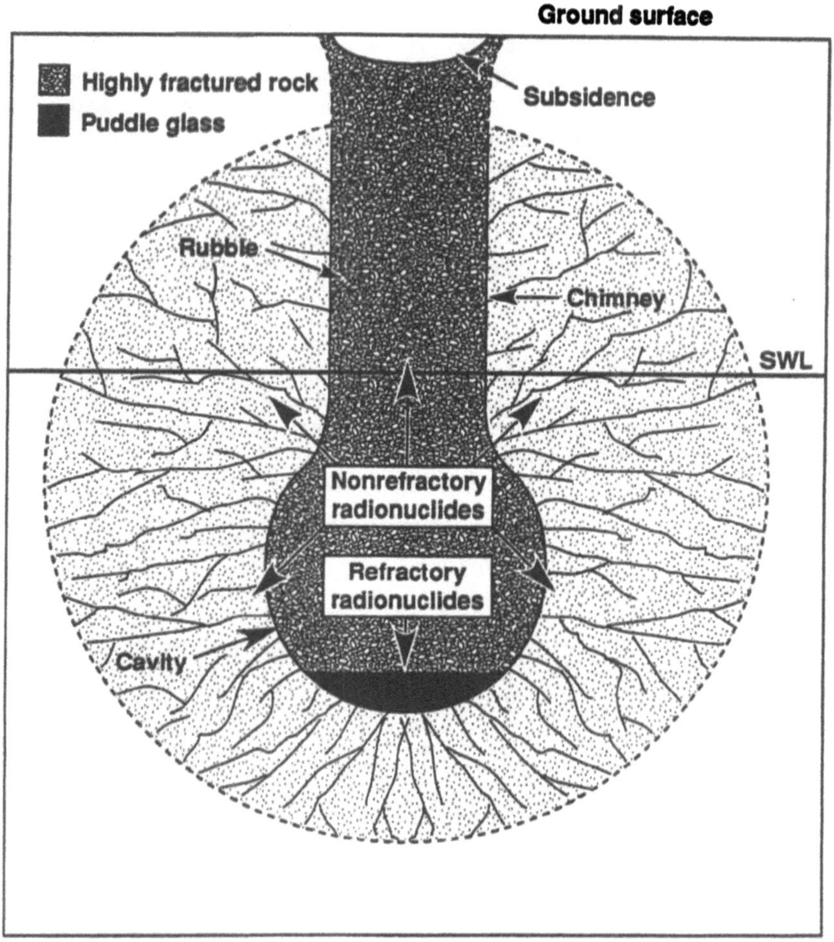


*Figure 2.* Cavity formation and other relevant geophysical features following detonation of a nuclear device below the standing-water level (SWL). From [17].

A more significant problem relates to the residual concentrations of radionuclides in surface soil. By 1970 the Nevada Applied Ecology Group (NAEG) had been formed at the NTS as a result of management's desire to respond to the goals of newly enacted legislation concerning protection of the environment [18]. The charter of the NAEG was broad and encompassed all ecological and environmental problems. Emphasis soon focused on radioecology, and, reflecting the concerns of the time, the programme focused even more narrowly on the radioecology of plutonium. As the programme matured, the focus broadened to include all other man-made radionuclides in surface soil. The overall goal was always to establish the scientific basis for the prediction of radiation dose from NTS operations to nearby residents, current on-site workers, and possibly future on-site residents.

One of the major accomplishments of the NAEG was to define the radioecological behaviour of plutonium in a natural ecosystem enhanced by the addition of grazing cattle; this study included an examination of the plutonium that was accumulated by cattle under such conditions [19, 20]. One of the features of this study was the demonstration that the contamination of vegetation in such environments, while significant, did not occur through the usual soil-root pathway, but by resuspension and subsequent redeposition on plant surfaces. This process was well illustrated by a comparison of vegetation-to-soil concentration ratios [21] as measured in the field, as opposed to measurements taken under glasshouse conditions; such data are summarized in Table 3.

TABLE 3. Summary [21] of data on the contamination of vegetation by plutonium in soil under field and glasshouse conditions. The data indicate that contamination under field conditions occurs not by the usual soil–root pathway, but by resuspension and redeposition. The term CR indicates the ratio of concentration of Pu (dry weight) in vegetation to that in soil. The Nellis Air Force Base Range (NAFBR) is the place where several plutonium-dispersion experiments took place

| Soil source | CR, field conditions | CR, glasshouse conditions |
|---|---|---|
| NTS Area 11B | 0.013–0.16 | 0.00015 |
| NTS Area 11C | 0.045–0.34 | 0.00018 |
| NTS Area 11D | 0.027–0.17 | 0.0011 |
| NAFBR Area 13 | 0.078–0.44 | 0.00011 |
| NAFBR Double Tracks | 0.011–0.094 | 0.00026 |
| NAFBR Clean Slate 1 | 0.0074–0.042 | 0.00043 |
| NAFBR Clean Slate 2 | 0.014–0.092 | 0.00076 |
| NAFBR Clean Slate 3 | 0.039–0.052 | 0.00055 |

The resuspension of Pu-contaminated soil particles was also measured directly. Out of these measurements were developed new models of resuspension that included a more realistic version of the time-dependent resuspension factor and the definition of the mass-loading approach [22].

Of particular interest were the accomplishments of a subprogramme of the NAEG, the Radionuclide Inventory and Distribution Programme (RIDP). This programme had the goal of providing a complete inventory and distribution of all man-made radionuclides in the surface soil of the NTS. This ambitious goal was accomplished by the use of a combination of the following techniques:

- Aerial surveys to establish the areas of significant radionuclide deposition;
- Soil sampling and analysis to establish the distribution with depth of the various radionuclides and to measure the concentration of non-gamma-emitting radionuclides; and
- Field-gamma spectrometry for quantitative evaluation of the concentrations of gamma-emitting radionuclides.

The scope of the programme was unique in that a complete radiological survey of such a large area had never been attempted before; the data collection took five years of nearly continuous effort, and analysis of the data was not completed for another five years. Results [23] of this inventory effort are summarised in Table 4.

TABLE 4. Inventory of the more significant radionuclides on the NTS in 1983 [23]

| Radionuclide | Inventory (TBq) |
|---|---|
| $^{241}Am$ | 5.9 |
| $^{238}Pu$ | 4.8 |
| $^{239+240}Pu$ | 36 |
| $^{60}Co$ | 3.3 |
| $^{137}Cs$ | 13 |
| $^{90}Sr$ | 17 |
| $^{152}Eu$ | 6.7 |
| $^{154}Eu$ | 1.3 |
| $^{155}Eu$ | 0.7 |

## 3. Radiation Dose to Members of the General Population

The most significant concern relating to the conduct of nuclear tests is the radiation dose that was received by members of the general population living downwind of the tests. Once the dose is known it is possible to estimate health effects, if the linear, no-threshold response is assumed.

### 3.1. PACIFIC ATOLLS

During the conduct of the early tests in the Marshall Islands, it was not yet realised how substantial the radiation doses might be from the larger tests. Thus, in 1954 with the detonation of the Bravo test on 1 March on Bikini, substantial radiation exposures occurred. During the course of a few days, 64 Marshallese were evacuated from Rongelap Atoll, 18 from Ailinginae Atoll, and 159 from Utirik Atoll. Among the evacuated persons were three pregnant women from Rongelap, one from Ailinginae, and eight from Utirik. In addition 28 military weather observers on Rongerik Atoll were evacuated. The estimated doses that these 281 persons received are indicated in Table 5 [24, 25]. In addition, 23 men aboard the Japanese fishing vessel, the *Fukuru Maru*, were exposed from being in the path of high fallout. On the basis of their symptoms, they received a dose of approximately 2 Gy. Thus, the highly exposed population received a collective effective dose of approximately 200 person Sv. A larger number of persons living on other atolls would have received additional, but much smaller, individual doses. As far as known to the author, there has not been an estimate of the total collective dose to this lesser-exposed population, either from the Bravo Shot or from the total number of explosions.

TABLE 5. Doses to the persons exposed to substantial levels of radiation following the Bravo Shot on 1 March 1954. Data are primarily from reference [24] with a correction for the persons exposed on Ailinginae from reference [25]

| Location | Number | Evacuation time after shot (h) | Total gamma dose in air (Gy) |
|---|---|---|---|
| Rongelap Atoll | 67 | 50–51 | 1.9 |
| Ailinginae Atoll | 19 | 58 | 1.1 |
| Utirik Atoll | 167 | 55–78 | 0.11 |
| Rongerik Atoll (USA) | 28 | 28.5–34 | 0.78 |

## 3.2. NEVADA TEST SITE

In contrast with the situation at the Marshall atolls, there never was a serious exposure situation arising from tests at the NTS, although there were a few isolated evacuations of personnel at small ranches and construction camps. There was also one instance (Shot Harry, a 32-kt shot on 19 May 1953) when a major highway was closed and persons were asked to stay indoors. Individual exposures were assessed for every shot, and a record of cumulative exposures was kept by personnel of the National Weather Nuclear Support Office in Las Vegas. Their map [13] of cumulative individual exposures is shown in Figure 3. As shown there, the highest contour is of 10 R, which would be equivalent to a dose in air of 0.09 Gy. In fact, very few people received a dose this high. The 1.0-R contour encloses a fairly substantial number of communities.

Because of public concern over the exposures that had been received, the US Department of Energy established a major dose-reconstruction effort known as the Off-Site Radiation Exposure Review Project (ORERP) in 1979. The goal of this project was to recalculate the doses from external gamma exposure and to calculate comprehensively for the first time the dose from the ingestion of radionuclides with food [26]. The assessment domain of the ORERP was initially that as shown in the hatched section (Phase I) of Figure 4, but was extended to the larger area indicated by the diagonal stripes (Phase II) also in Figure 4. This project was officially concluded in 1987.

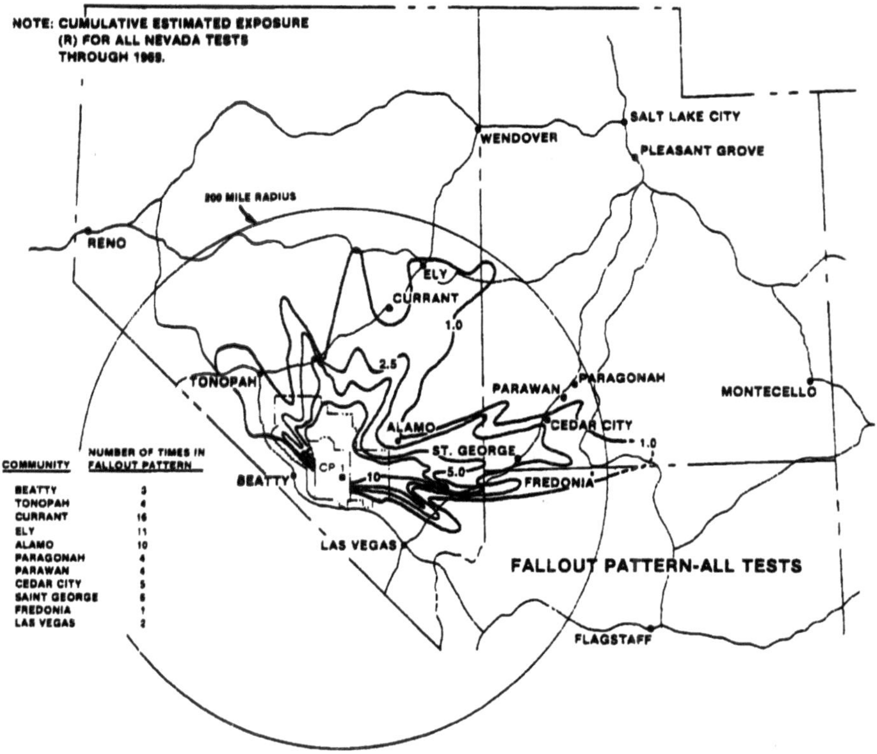


*Figure 3.* Map of estimated cumulative exposure as developed by the US Weather Nuclear Support Office in Las Vegas, NV [13].

The methods of dose reconstruction used for the ORERP are discussed elsewhere [27-31]. The main results for the Phase I and II areas were a collective dose from external gamma exposure of 12 000 person-Gy [32]; the collective organ doses due to the ingestion of radionuclides with food are indicated in Figure 5 [31]. The only organ for which the dose is higher for ingestion as compared to external gamma exposure is the thyroid, which has a collective organ dose of 140 000 thyroid-Gy.

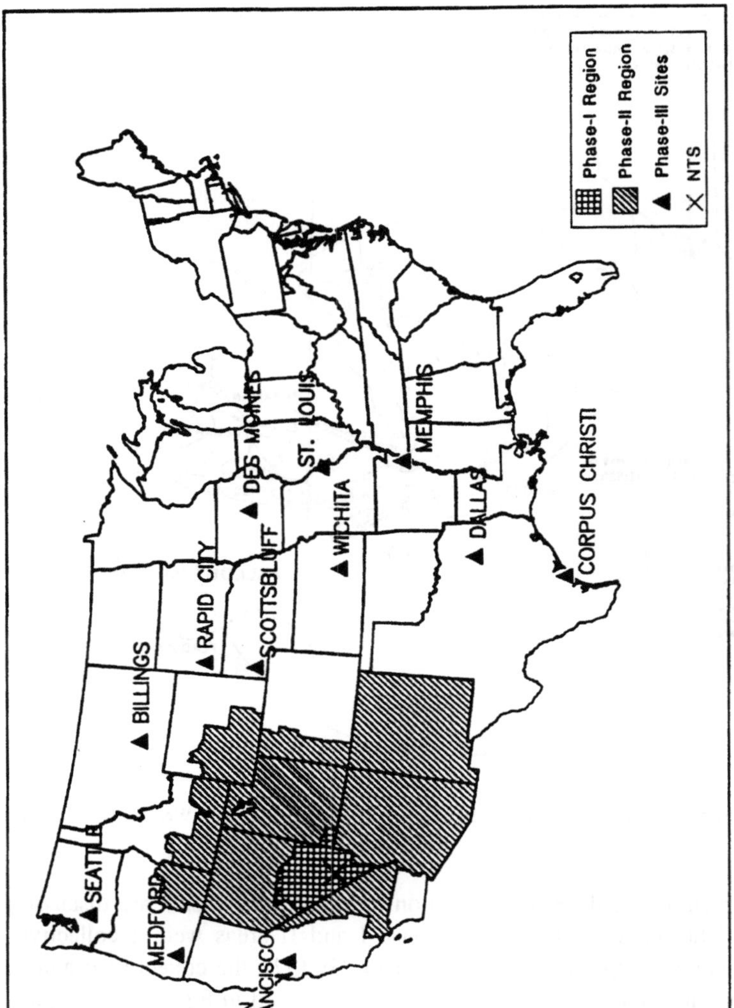


*Figure 4.* The assessment domain of the ORERP [28].

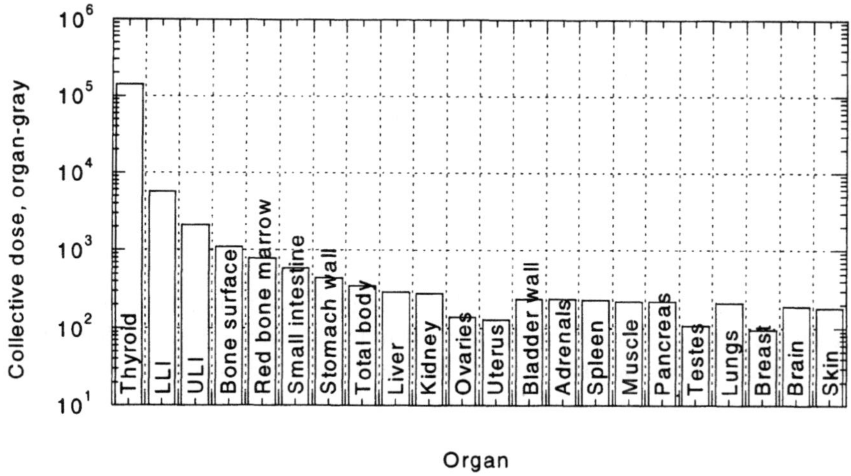


*Figure 5.* Collective organ doses in the Phase I and Phase II areas of the ORERP. Data from [31].

## 4. Health Effects

If health effects are severe enough to be observed directly, they can be tabulated and enumerated. A more frequent result for radiation exposure is that a large population receives a fairly small dose; under such conditions, it is usually extremely difficult to find radiogenic effects, owing to the large background incidence of long-term stochastic effects (such as nonfatal or fatal cancer). In such cases, the primary health effects can be estimated with the assumption of a linear, no-threshold, dose-effect response. It is usual to estimate the induction of fatal cancers by using the International Commission on Radiological Protection (ICRP) risk estimator of 0.05 fatal cancers per sievert of effective dose [33] or per gray of whole body absorbed dose. Such estimates can be compared to the number of persons expected to die of fatal cancer from normal causes, which is estimated to be 20% of the population.

### 4.1. PACIFIC ATOLLS

Owing to the exposures from the Bravo Shot on 1 March 1954, 304 persons were exposed to high doses; many of them experienced radiation sickness. One of the Japanese fisherman died from serum hepatitis within a few months of exposure. His death was likely the result of an infection received during the course of many blood transfusions [34].

In 1963 two 12-year old Rongelap females were found to have thyroid nodules, which were removed and diagnosed as adenomatous hyperplasia [25]. A follicular carcinoma was found in 1969 in an adult female from Utirik [25]. There have now been observed 7 adenomas, 9 papillary cancers, 1 follicular cancer, and 7 occult cancers in the 253 exposed Marshallese as compared to 1, 2, 0, and 2 cases in 227 members of a comparison group [25]. No mortality from thyroid carcinoma has been observed in the group of exposed Marshallese [25].

The number of radiogenic fatal cancers that might be induced in the 304 highly exposed persons is 10, as compared to 60 cases due to normal causes.

## 4.2. NEVADA TEST SITE

There are no well-documented cases of acute radiation sickness from the tests conducted at NTS. A variety of studies has been performed to investigate whether an increased incidence of cancer can be detected in the off-site population. Stevens *et al.* [35] conducted a study of 1177 cases of leukaemia and 5330 controls who were born before 1 November 1958 and died between 1 January 1952 and 31 December 1981 while residents of Utah. No dose-related effect was found for all forms of leukaemia combined, but an effect was found for a defined subgroup: acute leukaemias discovered from 1952–1963 among those younger than 20 years at exposure. (This association was based upon five cases of acute leukaemia.)

The thyroid study reported by Kerber *et al.* [36] examined a cohort originally defined in the late 1960s of 4818 children; 2687 of these children were enrolled in grades 5 through 12 in 1965–1970 in Washington County, Utah, and in Lincoln County, Nevada, two counties judged to be heavily impacted by fallout; the other 2131 children were of a similar age and lived in Graham County, Arizona, a county judged suitable for a control. Reports of the original study indicated that two carcinomas were found, one in Nevada and one in Arizona; evidence for any radiation effect on thyroid disease was judged to be statistically insignificant [37]. Some members (2579; limited to white non-Hispanics still living in the three-state area) of the original cohort were re-examined in 1985 and 1986; interviews were also conducted of the parents of all but 53 subjects in order to provide input data for calculation of individual-thyroid doses for the subjects in all three states [38]. A total of 2473 persons were included in the analysis of the data for the period prevalence of 1965 through 1986 [36]. There were 56 subjects with thyroid nodules, 19 with neoplasms (adenoma plus carcinoma), and eight with papillary carcinoma (no follicular carcinomas occurred among the analysed subjects). A statistically significant positive dose-response trend was observed for neoplasms ($p = 0.019$), but not for thyroid nodules ($p = 0.16$) or for carcinomas ($p = 0.096$).

Given the 12,000 person-gray collective dose to the off-site population, it is possible to calculate that 600 cases of fatal cancer might result. This dose was accrued by 11 275 000 persons, so the background number of fatal cancers to be expected is approximately 2,300,000. It would be extremely difficult to detect whether such an increase actually occurred against this high background.

## 5. References

1. US Department of Energy (1994) United States Nuclear Tests, July 1945 through September 1992, US Department of Energy Nevada Operations Office, Las Vegas, NV, DOE/NV-209 (Revision 14).
2. Webb, J.S. (1949) The fogging of photographic film by radioactive contaminants in cardboard packaging material, *Phys. Rev.* **76,** 375-380.
3. Simon, S.L. and Vetter, R.J. (1997) Consequences of nuclear testing in the Marshall Islands. *Health Phys.* **73,** No. 1.
4. Glasstone, S. and Dolan, P.J. (1977) The Effects of Nuclear Weapons, US Department of Defense and US Department of Energy, Washington, DC, 3rd edition.
5. Hicks, H.G. (1982) Calculation of the concentration of any radionuclide deposited on the ground by offsite fallout from a nuclear detonation, *Health Phys.* **42,** 585-600.
6. Hicks, H.G. (1990) Additional calculations of radionuclide production following nuclear explosions and Pu isotopic ratios for Nevada Test Site events, *Health Phys.* **59,** 515-523.
7. Henderson, R.W. (1991) Approximation of the Decay of Fission and Activation Product Mixtures, Los Alamos National Laboratory, Los Alamos, NM, LA-11968-MS.
8. Simon, S.L. (1997) A brief history of people and events related to atomic weapons testing in the Marshall Islands, *Health Phys.* **73,** 5-20.
9. Niedenthal, J. (1997) A history of the people of Bikini following nuclear weapons testing in the Marshall Islands: with recollections and views of elders of Bikini Atoll, *Health Phys.* **73,** 28-36.
10. Robison, W.L., Noshkin, V.E., Conrado, C.L., Eagle, R.J., Brunk, J.L., Jokela, T.A., Mount, M.E., Phillips, W.A., Stoker, A.C., Stuart, M.L. and Wong, K.M. (1997) The Northern Marshall Islands radiological survey: data and dose assessments, *Health Phys.* **73,** 37-48.
11. Robison, W.L., Bogen, K.T. and Conrado, C.L. (1997) An updated dose assessment for resettlement options at Bikini Atoll—a U.S. nuclear test site, *Health Phys.* **73,** 100-114.
12. Noshkin, V.E., Robison, W.L., Wong, K.M., Brunk, J.L., Eagle, R.J. and Jones, H.E. (1997) Past and present levels of some radionuclides in fish from Bikini and Enewetak Atolls, *Health Phys.* **73,** 49-65.
13. Friesen, H.N. (1985) A Perspective on Atmospheric Nuclear Tests in Nevada, US Department of Energy Nevada Operations Office, Las Vegas, NV, NVO-296.
14. Davis, M.G., Flotard, R.D., Fontana, C.A., Hennessey, P.A., Maunu, H.K., Mouck, T.L., Mullen, A.A. and Sells, M.D. (1999) Offsite Environmental Monitoring Report: Radiation Monitoring Around United States Nuclear Test Areas, Calendar Year 1997, US Environmental Protection Agency Radiation and Indoor Environments National Laboratory, Las Vegas, NV, EPA-402-R-98-013.
15. Anspaugh, L.R., Black, S.C., Costa, C.F., Elle, D.R., Essington, E.H., Gilbert, R.O., Gonzalez, D.A., Hunter, R.B., McArthur, R.D., Medica, P.A., O'Farrell, T.P., Patton, S.E., Romney, E.M., Shinn, J.H. and Thompson, C.B. (1990) Radiation-related monitoring and environmental research at the Nevada Test Site, in *Environmental Monitoring, Restoration, and Assessment: What Have We Learned?* Battelle Press, Columbus, OH, pp. 159-167.
16. Marsh, K.V. (1992) Appendix A. Tests fired near or below the water table, in *Hydrology and Radionuclide Migration Programme 1987 Progress Report,* Lawrence Livermore National Laboratory, Livermore, CA, UCRL-53779-89.
17. Daniels, J.I., Andricevic, R., Anspaugh, L.R. and Jacobson, R.L. (1993) Risk-based screening analysis of ground water contaminated by radionuclides introduced at the Nevada Test Site (NTS) in *Pilot Study Risk Assessment for Selected Problems at the Nevada Test Site (NTS),* Lawrence Livermore National Laboratory, Livermore, CA, UCRL-LR-113891, pp. 69-97.
18. Anspaugh, L.R. (1992) An historical perspective of the Nevada Applied Ecology Group, in *Summary of the Nevada Applied Ecology Group and Correlative Programmes,* US Department of Energy Nevada Operations Office, Las Vegas, NV, DOE/NV-357, pp. 97-116.

19. Gilbert, R.O., Engel, D.W., Smith, D.D., Shinn, J.H., Anspaugh, L.R. and Eisele, G.R. (1988) Transfer of aged Pu to cattle grazing on a contaminated environment, *Health Phys.* **54**, 323-335.
20. Gilbert, R.O., Engel, D.W., and Anspaugh, L.R. (1989) Transfer of aged $^{239+240}$Pu, $^{238}$Pu, $^{241}$Am, and $^{137}$Cs to cattle grazing a contaminated arid environment, *Sci. Tot. Environ.* **85,** 53-62.
21. Romney, E.M. and Wallace, A. (1977) Plutonium contamination of vegetation in dusty field environments, in *Transuranics in Natural Environments,* US Department of Energy Nevada Operations Office, Las Vegas, NV, NVO-178, pp. 287-302.
22. Anspaugh, L.R., Shinn, J.H., Phelps, P.L. and Kennedy, N.C. (1975) Resuspension and redistribution of plutonium in soils, *Health Phys.* **29,** 571-582.
23. McArthur, R.D. (1991) Radionuclides in Surface Soil at the Nevada Test Site, US Department of Energy Nevada Operations Office, Las Vegas, NV, DOE/NV/10845-02.
24. Cronkite, E.P., Conard, R.A. and Bond, V.P. (1997) Historical events associated with fallout from Bravo Shot—Operation Castle and 25 y of medical findings, *Health Phys.* **73,** 176-186.
25. Howard, J.E., Vaswani, A. and Heotis, P. (1997) Thyroid disease among the Rongelap and Utirik population—An update, *Health Phys.* **73,** 190-198.
26. Church, B.W., Wheeler, D.L., Campbell, C.M., Nutley, R.V. and Anspaugh, L.R. (1990) Overview of the Department of Energy's Off-Site Radiation Exposure Review Project (ORERP), *Health Phys.* **59,** 503-510.
27. Anspaugh, L.R. (2000) Technical basis of reconstruction of dose from tests of nuclear weapons, *this volume.*
28. Beck, H.L. and Anspaugh, L.R. (1991) Development of the County Database: Estimates of exposure rates and times of arrival of fallout in the ORERP Phase-II Area. Comparison with cumulative deposition-density estimates based on analyses of retrospective and historical soil samples, US Department of Energy Nevada Operations Office, Las Vegas, NV, DOE/NV–320.
29. Henderson, R.W. and Smale, R.F. (1990) External exposure estimates for individuals near the Nevada Test Site, *Health Phys.* **59,** 715–721.
30. Ng, Y.C., Anspaugh, L.R. and Cederwall, R.T. (1990) ORERP internal dose estimates for individuals, *Health Phys.* **59,** 693-713.
31. Whicker, F.W., Kirchner, T.B., Anspaugh, L.R. and Ng, Y.C. (1996) Ingestion of Nevada Test Site fallout: internal dose estimates, *Health Phys.* **71**, 477-486.
32. Henderson, R.W. and Smale, R.F. (1992) Summary of Collective Dose from External Exposure, Los Alamos National Laboratory, Los Alamos, NM, Report submitted to the Co-ordination and Information Centre, Las Vegas, NV.
33. International Commission on Radiological Protection (1991) 1990 Recommendations of the International Commission on Radiological Protection, Pergamon Press, Oxford, ICRP Publication 60.
34. Eisenbud, M. (1997) Monitoring distant fallout: the role of the Atomic Energy Commission Health and Safety Laboratory during the Pacific tests, with special attention to the events following Bravo, *Health Phys.* **73,** 21-27.
35. Stevens, W., Thomas, D.C., Lyon, J.L., Till, J.E., Kerber, R.A., Simon, S.L., Lloyd, R.D., Elghany, N.A. and Preston-Martin, S. (1991) Leukemia in Utah and radioactive fallout from the Nevada Test Site, *J. Am. Med. Assoc.* **264,** 585-591.
36. Kerber, R.A., Till, J.E., Simon, S.L., Lyon, J.L., Thomas, D.C., Preston-Martin, S., Rallison, M.L., Lloyd, R.D. and Stevens, W. (1993) A cohort study of thyroid disease in relation to fallout from nuclear weapons testing, *J. Am. Med. Assoc.* **270,** 2076-2082.
37. Rallison, M.L., Dobyns, B.M., Keating, F.R., Rall, J.E. and Tyler, F.H. (1975) Thyroid nodularity in children, *J. Am. Med. Assoc.* **233,** 1069-1072.
38. Till, J.E., Simon, S.L., Kerber, R., Lloyd, R.D., Stevens, W., Thomas, D.C., Lyon, J.L. and Preston-Martin, S. (1995) The Utah thyroid-cohort study: analysis of the dosimetry results, *Health Phys.* **68**, 472–483.

# HYDROGEOLOGICAL INVESTIGATIONS AND RADIONUCLIDE MIGRATION IN GROUNDWATER AT THE NEVADA TEST SITE

W. L. HAWKINS
*Earth and Environmental Science Division*

J. L. THOMPSON, D. L. FINNEGAN
*Chemical Science and Technology Division,*
*Los Alamos National Laboratory, Los Alamos, NM 87545*

**Abstract**

Although the United States stopped nuclear testing in 1992, in the preceding three decades more than 800 nuclear devices were detonated underground at the Nevada Test Site (NTS) in southern Nevada. The majority of the detonation points of those tests are located above the water table, primarily in alluvium and volcanic rocks. In the main testing areas, the water table is 450–700 m below the surface. Pre- and post-event geologic investigations were conducted for each test and longer-term studies are underway to assess the regional environmental impact. Studies now in progress have not identified any impact on the regional groundwater flow system from testing, but local effects have been discovered.

In some areas where several tests have been conducted below the water table, flow-through and straddle-packer testing have revealed unexpectedly high hydraulic pressures at depth. The movement of radionuclides underground in the vicinity (~1 km) of individual test locations has been studied and has identified several mechanisms for this movement. One mechanism is prompt (within one second) injection of volatilised radioactive material through explosion-produced fractures and may transport some refractory elements as well. A second mechanism is gas pumping through pre-existing or explosion fractures. Only gases or easily volatilised elements are involved, and this process ceases within a few minutes when temperatures and pressures reduce toward normal values. (Of course, gaseous elements may continue to diffuse through the geologic media.) A third mechanism is movement with groundwater. Radionuclides that dissolve in water will be transported away from the test location driven by the hydraulic gradients. Radionuclides that are insoluble in water may move also if they form colloids or can sorb on naturally occurring colloids.

At the NTS fracture, injection over distances of a few hundred meters has been observed. Gas pumping normally moves volatile radionuclides a few tens of meters from the nuclear test cavity. There are no known bounds on how far nonsorbing species such as tritium (as tritiated water) or dissolved krypton can move with groundwater.



*S.S. Hecker et al. (eds.), Nuclear Physical Methods in Radioecological Investigations of Nuclear Test Sites*, 61-74.
© 2000 *Kluwer Academic Publishers. Printed in the Netherlands.*

Movement of plutonium and americium has been observed over distances of a kilometer, but the mechanism by which this movement occurred has not yet been determined.

## 1. Introduction

The Nevada Test Site (NTS), located in the high desert of the southern Great Basin approximately 100 km northwest of Las Vegas, Nevada, (Fig. 1) occupies an area of 3500 $km^2$ with elevations ranging from 850 to 2340 m above sea level. Annual precipitation is 0.1 to 0.3 m and temperatures vary between -26 and 43 °C. Except for a few small springs, there is no permanent surface water on the NTS.

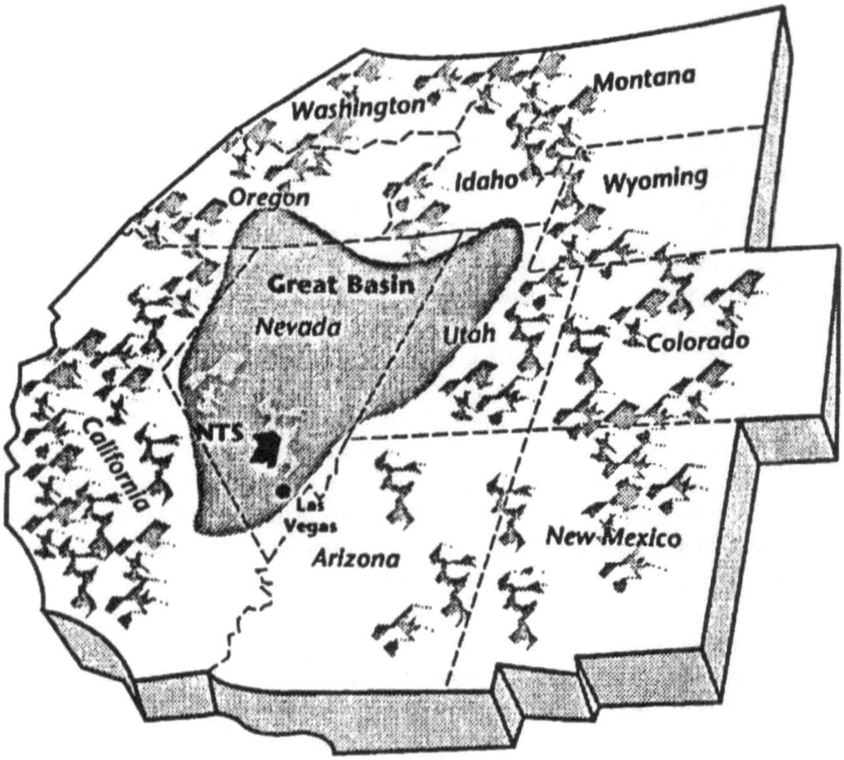


*Figure 1.* Location of the Nevada Test Site within the southern Great Basin of the western United States.

The surface geology of the NTS (Fig. 2) is approximately 40% alluvium-filled basins, 20% Paleozoic and uppermost Precambrian sedimentary outcrops, and the remainder is Tertiary volcanics with a few Mesozoic intrusive masses [1,2]. In this structurally-complex region of crustal extension, there are numerous basin-range style normal faults, low-angle thrust faults, and a few prominent strike-slip faults [3].

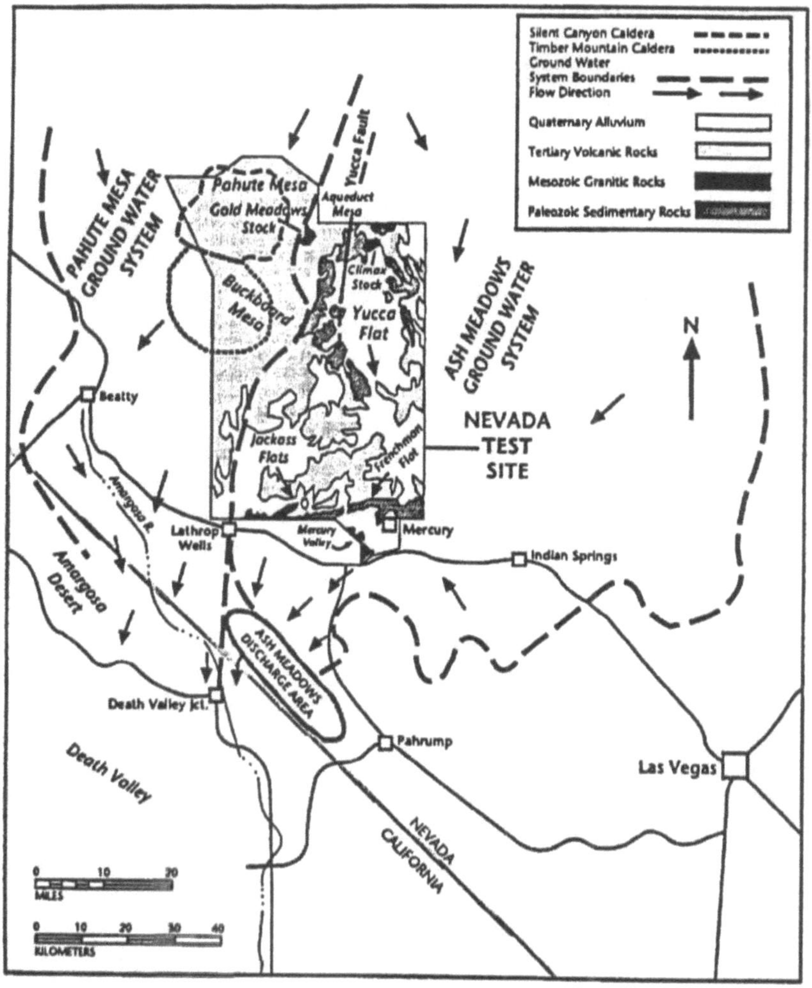


*Figure 2.* Map of the Nevada Test Site geology and regional groundwater-flow system.

Groundwater beneath the NTS moves to the south-southwest (Fig. 2). Three hydrogeologic units control the movement of groundwater: the lower clastic confining unit, the lower carbonate aquifer, and the tuff confining unit [4]. Perched groundwater is present in some areas. Underground nuclear detonations are primarily conducted in unsaturated alluvium, tuffs, and lavas of the northern and eastern portions of the NTS, where the water table is 450 - 700 m below the ground surface [5].

Since 1976 the distribution and movement of radionuclides associated with underground nuclear tests at the NTS have been studied. Generally work is focused on radionuclides found within a few hundred meters of the cavity, although radionuclides associated with particular tests have been discovered at distances of 1.3 km.

## 2. Dispersal of Radionuclides

As a result of the formation and collapse of the cavity following the nuclear explosion, radionuclides are deposited in heterogeneous concentrations within a few cavity radii of the detonation point. The major fraction of radioactivity is found in the "melt glass," which forms at the bottom of the cavity and is mixed with rubble as the cavity collapses.

A small fraction of the radionuclides may be expelled from the cavity through fractures in the surrounding rock. This process, called "prompt injection" occurs less than 1 s after the detonation and has been observed to transport both volatile and refractory materials distances of hundreds of meters. However, it does not appear that prompt injection occurs at every test site, or that this mechanism is responsible for dispersing large fractions of the radioactive material associated with nuclear tests [6].

Radionuclides that are gases or volatiles may move through pre-existing or explosion-generated fractures around the cavity within a few minutes after the event while temperatures and pressures are elevated. This movement, which generally results in dispersal of volatile materials a few tens of meters around the cavity, is an important mechanism for dispersing $^{3}H$, $^{85}Kr$, and $^{137}Cs$, the latter having gaseous precursors.

## 3. Migration of Radionuclides Through the Subsurface

Studies to date indicate that moving groundwater is required for further dispersal of radionuclides from the site of a nuclear test after pressure and temperature conditions are returned to near-ambient values. A number of sites have been studied where tests were conducted below the water table and where the sites were filled again with water post-test [7,8]. Several of these sites were under water for over ten years. At those sites where there was little hydraulic gradient and the returned water was essentially stagnant, little movement of radionuclides occurred. Even tritium (in the form of water) seemed to remain in place with little dispersion or mixing. However, in situations where there was groundwater flow because of hydraulic gradients and good hydraulic conductivity, many radionuclides were mobilised. The controlling mechanisms (other than groundwater flow) have to do with the deposition of the radionuclides (e.g., surface deposited or occluded in a solid matrix), with sorption properties of the dissolved species, with the groundwater chemistry such as ionic strength and oxidation potential, and with the speciation of the radionuclides. Some radionuclides may be transported in the form of colloids [9]. Research is continuing so that this mechanism for radionuclide transport may be better understood.

It is not known how far or how fast radionuclides may ultimately move underground at the NTS, for the observation time there is only a little more than 30 years. However, the data so far suggest that movement over distances of kilometers is relatively rare and probably is confined to situations where there is substantial groundwater flow. There is almost no information about movement of radionuclides that were deposited above the water table in the vadose zone. The extent to which these materials could be mobilised by episodic events such as heavy rainfall is largely unknown, though there is no evidence for widespread dispersal through the vadose zone.

Current research is focused on: 1) modelling the data available from field and laboratory studies to develop predictive capabilities, and 2) understanding the mechanisms operative in the movement of very small concentrations of plutonium over distances of 1.3 km [10].

## 4. High Water Levels

Forty days after an underground nuclear detonation in May, 1962, the water level in a well located 305 m to the west of the event site was found to be about 17 m higher than the last pre-test measurement made in the summer of 1961 [11]. This water level had not fluctuated more than 0.6 m during the three preceding years. From the rate of decline records, a projection of two years was made for the fluid level to return to the pre-test level. After a 1963 underground nuclear detonation, measurements of fluid-level changes were made in four nearby holes [12]. This data also indicated immediate increase of water-table altitude and subsequent slow decline. It is yet to be determined if the elevated fluid levels are the result of actual elevation increases of the water table's mounding or slowly dissipating or of pore pressure increases in the tuff confining unit.

Ten of twenty-five emplacement sites, within or near a geologically well-characterised area in the centre of Yucca Flat, have required remedial actions to make them acceptable for event execution (Fig. 3). At five the large diameter (>1.5 m) emplacement hole had water entering the hole over the top of the liner. (To provide a dry environment for the experiment, liners are installed in emplacement holes from above what is believed to be the water table to below the detonation point.) Three sites in the emplacement hole had liners that leaked. All the leaks were within 20 m of the bottom of the liner (~600 m depth). Packer tests in exploratory holes at the two sites indicated formation pressures that exceeded the design limits of their liners.

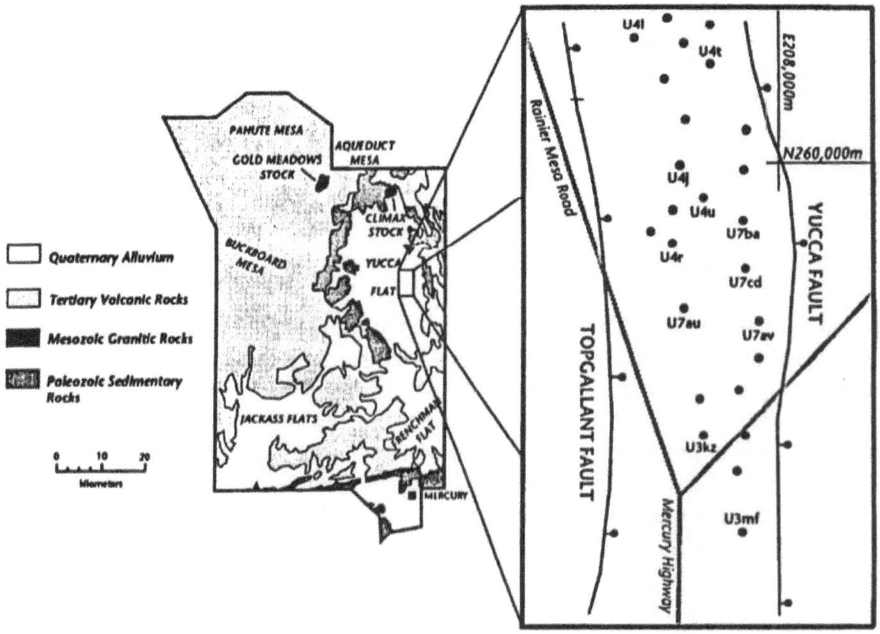


*Figure 3.* Underground nuclear test locations in a portion of Yucca Flat. Labelled sites required remedial actions owing to the elevated water table and/or failed liners.

## 4.1. U3kz EXPLORATION

At the U3kz site, the geologic setting is well understood (Figs. 4(a) and 4(b)). The water table was estimated to be at a depth of 500 m. In August of 1984, a 2.44 m diameter emplacement hole was drilled and lined with a 1.88 m diameter liner from a depth of 449 m to 670 m. Water was discovered rising in the liner. The source, which was a leak near the liner bottom, was providing inflow at a fairly consistent rate of rise of 1.5 m per day. After several months, the water level had stabilised at a depth of approximately 435 m or 14 m above the top of the liner and 65 m higher than the predicted water table [13]. Elevated levels of tritium ($5 \times 10^4$ pCi/ml) and trace amounts of fission products were discovered in fluid samples. The source for the fission products is the event conducted in U7aq, 337 m to the north.

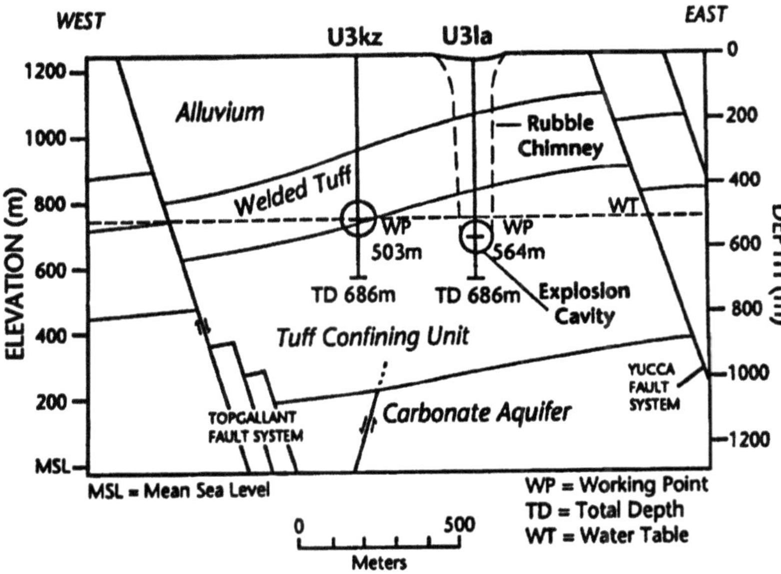


*Figure 4(a).* East-to-west geologic cross-section through U3kz site.

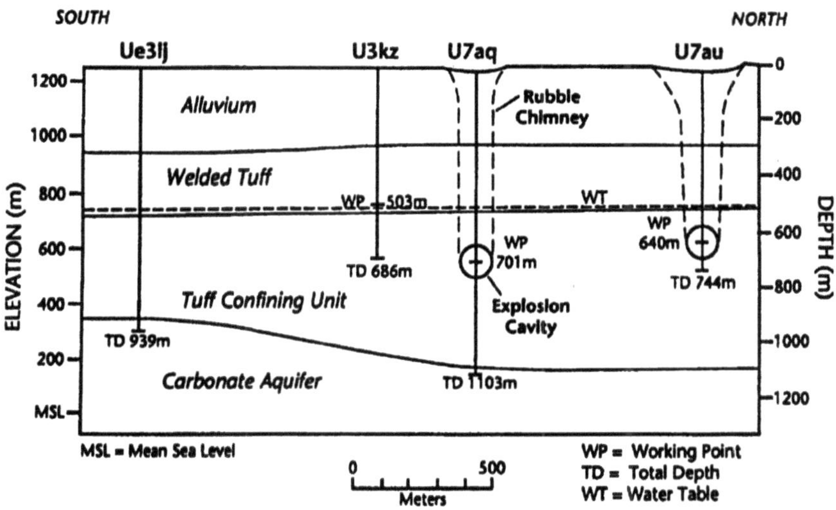


*Figure 4(b).* North-to-south geologic cross-section through U3kz site.

An exploratory hole (Ue3e) was drilled approximately 65 m to the north of U3kz and in the direction of U7aq. This Ue3e hole was drilled in three phases with a waiting period between each phase to monitor the fluid level in the open hole. The hole was drilled to a depth of 520 m for phase one; 642 m for phase two; and 735 m for phase three. Owing to loss of hole depth following phase three, the hole had to be cleaned out and deepened to 765 m to ensure adequate depth for data acquisition. After the drilling of phases one and three and the final deepening, the depth-to-fluid in the open hole stabilised within a few meters of 470 m (Fig. 5). The shallower, apparently stabilised, depth-to-fluid following phase two is unexplained and may be related to a corresponding loss of total depth.

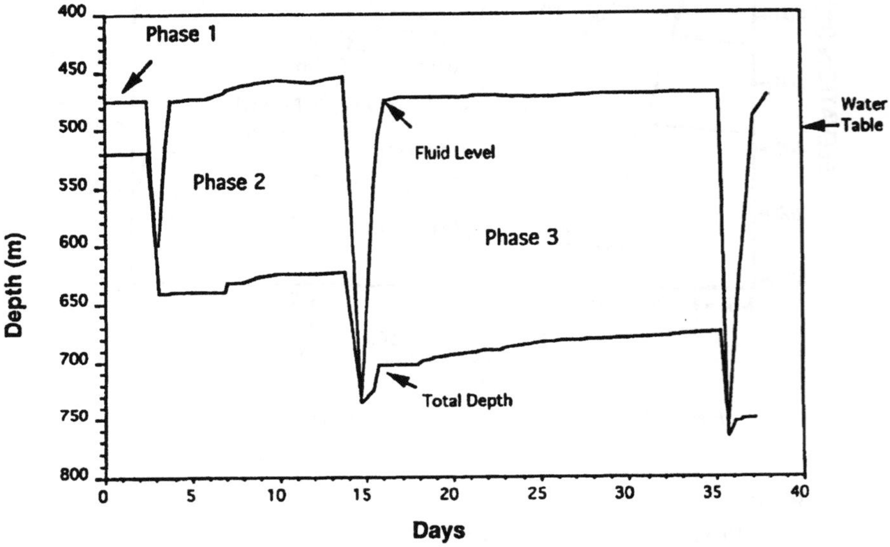


*Figure 5*. Hole Ue3e fluid-level and total-depth history during staged drilling.

Fluid samples monitored during the drilling indicated elevated levels of tritium beginning at a depth of 661 m. A natural gamma log showed a spike of activity at 659 m depth and a sidewall sample taken in that interval contained a trace of $^{137}Cs$.

A flow-through packer was set in the Ue3e open hole at a depth of 646 m. Fluid rose in the drill pipe running through the packer to a measured depth of 255 m, more than 250 m above the predicted water table (Fig. 6). The fluid level had not stabilised after more than ten days. The fluid in the annulus stabilised within 24 hours at a depth of 470 m. The consistency in the depths of the fluid when the hole is open to the formation as opposed to the depth in the pipe during the flow through packer testing implies that the water is entering the hole at depth and flowing up and out of the hole in the unsaturated fractured tuffs.

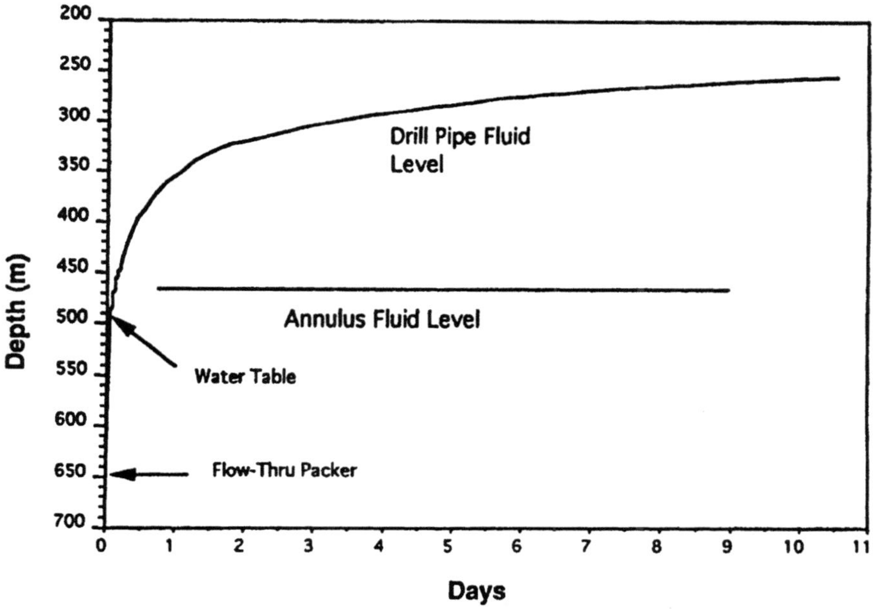


*Figure 6.* Fluid-level history in the drill pipe and annulus during the flow-thru packer test in Ue3e.

In March of 1987, six months after the execution of the U3kz event, another exploratory hole (Ue3e#2) was drilled 12 m to the north of Ue3e. A protective casing was installed, but not cemented, in the hole down to 594 m. Continuous core was recovered (~100% recovery) from 594 m to 723 m and the hole was geophysically logged. Tritium levels similar to those in U3kz were present; however, the zone of $^{137}Cs$ contamination found at 659 m in Ue3e was not penetrated. This finding implies that the contamination is not associated with a continuous horizontal feature such as a bedding plane, but in a small channel, perhaps associated with a high-angle fracture or fault that Ue3e#2 did not penetrate.

Five months after fluid sampling and geophysical logging, the fluid level in the Ue3e#2 casing was at a depth of 177 m. Shortly thereafter, the hole was lost during casing-removal operations. Because the casing was not cemented, there was communication between the fluid in the casing and the fluid in the annulus, so Ue3e#2 was considered an open hole. The height of the fluid in Ue3e#2 is surprising, especially when compared with the data from the open hole and the packer testing in Ue3e, 12 m to the south, which had much lower fluid levels in the open hole. However, Ue3e#2 was rotary drilled from the surface to 594 m with air foam, and the core drilling below 594 m required the use of a gel and polymer mud. These drilling-fluid additives, especially the latter, reduce fluid loss into the formation, perhaps resulting in the higher fluid level in the hole.

To increase the probability of intersecting the zone of caesium contamination, if it was associated with a vertical plane (high-angle fault or fracture), an angle hole through the zone was planned. Owing to hole stability problems, this effort had to be abandoned before it reached the point of interest.

A fourth exploration hole in the U3kz area was drilled (Ue3e#4) to again try to obtain rock samples from the zone where the $^{137}Cs$ was found in Ue3e and to install three peizometer tubes. Core recovery was poor through the zone of interest but sidewall samples did confirm the presence of $^{137}Cs$. Three peizometer tubes were installed to obtain fluid samples and monitor the depth to fluid. Fluid-level histories in the three tubes show an upward hydraulic gradient (Fig. 7).

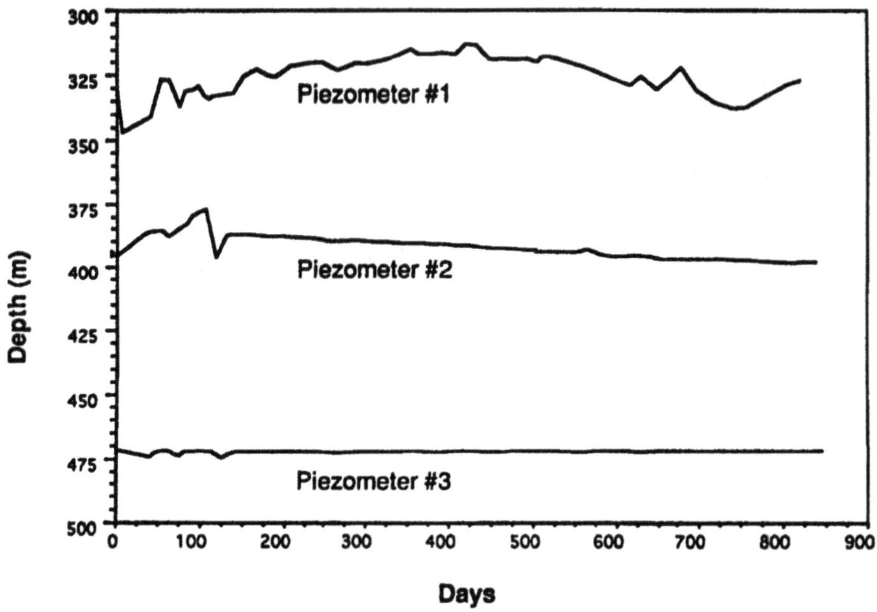


*Figure 7.* Fluid-level histories for piezometers set in hole Ue3e#4. Depths of slotted intervals are: Piezometer #1 = 666 m; Piezometer #2 = 586 m; and Piezometer #3 = 507 m. Water table was estimated at an elevation of 500 m.

Although the geology in the U3kz area is relatively uncomplicated, the hydrologic setting is quite complicated and underground testing has induced more complication by increasing the formation pressures and producing or reactivating fracture systems. The contamination from the previous test probably reached U3kz through high-angle fractures. Because the low-permeability zeolitized tuffs at this depth are highly sorptive of caesium [14], the most probable transport mechanism is dynamic gaseous injection. The highly mobile tritium most likely moved with the groundwater.

## 4.2 U4t EXPLORATION

The U4t hole was a 3.05-m diameter emplacement hole drilled to a total depth of 615.7 m in May of 1985. It was lined from 441 m to 601 m with a 1.88-m diameter liner for an estimated water table depth of 506 m [13]. During the month following installation of the liner, slowly rising fluid levels were detected in the liner. After dewatering of the liner, video inspection revealed a collapsed liner section with a leak about 5.5 m above the hemispherical liner bottom. To permit the use of the hole as an emplacement site, a 1.68-m diameter liner was cemented inside the original liner from a depth of 426 m to 599 m. The fluid level in U4t had not stabilised before the installation of the second liner. A ton of sodium bromide was placed below the second liner as a groundwater tracer for hydrologic and radionuclide migration studies. The tracer has not been detected in any drill holes.

Twelve meters to the west of U4t, satellite hole U4t #1 was drilled in April of 1986. Two months after the completion of this 0.31-m diameter, 640 m deep, uncased hole, the depth to fluid was 323 m. The fluid level stabilised after another two months at a depth of 315 m, or 191 m above the predicted static water level and over 100 m above the U4t liner top. During a borehole video inspection, fluid was observed seeping into U4t at a depth of 387.4 m through fractures in the welded tuff. This fluid is believed to have come from the satellite hole because five days after the fluid was removed from the satellite hole, the seepage in U4t ceased.

Drilling of Ue4t commenced in January of 1987, two months after the execution of the event in U4t. It was drilled 168 m south and 15 m west of U4t. The 0.25-m diameter hole was drilled in three phases by using air-foam to a total depth of 736 m and then left uncased. Fluid levels were monitored following each phase of drilling (Fig. 8). Data following phase one drilling were insufficient for projection of a stabilised fluid level; however, the stabilised phase-two fluid level was at a depth of approximately 320 m. The phase-three fluid level seems to have stabilised at a depth of about 265 m [15].

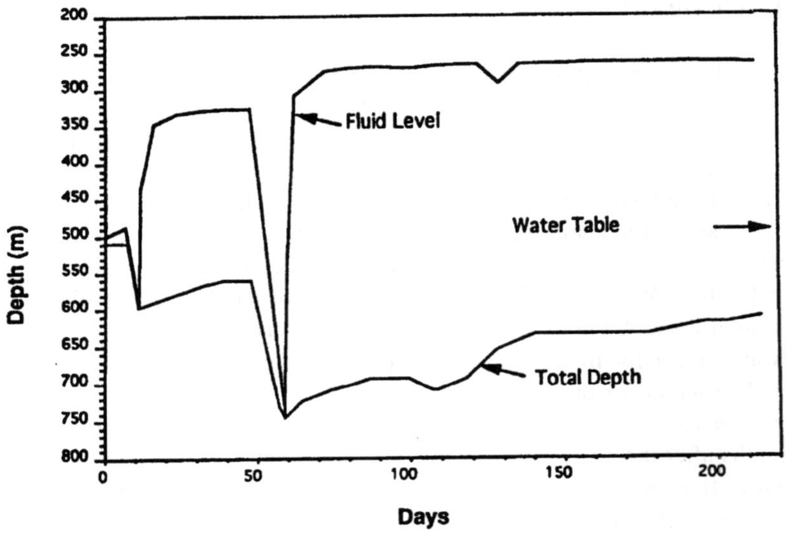


*Figure 8*. Hole Ue4t fluid-level and total-depth history during staged drilling.

Continuous water-level measurements by the United States Geological Survey (USGS) show that the level fluctuates about 1.5 m with corresponding changes in the atmospheric pressure. Analysis of temperature logs from Ue4t suggests fluid is entering the hole at depth, flowing up the hole, and exiting at several shallower depths [16].

Straddle-packer testing conducted in Ue4t during January, February, and May of 1989 confirms that there is a general upward hydraulic gradient. The lowest formation pressure (1979 kPa) was measured between the surface and a depth of 542 m, while the highest formation pressure (5205 kPa) was measured between 629 m and the bottom of the hole. The general upward pressure gradient is not evident between 542 m and 629 m depth as the formation pressures in this interval are between 3454 kPa and 3937 kPa. The depth of the water table in Ue4t was not determined owing to flow around the packers.

No elevated levels of tritium or fission products have been detected in Ue4t. Packer testing confirms the upward hydraulic gradient, which is probably a result of the underground tests in the area. Hydrologic investigations by the USGS in a hole drilled prior to underground nuclear testing in this area did not detect an upward gradient in the tuff confining unit [17].

## 5. Summary

Depth-to-fluid measurements and packer tests in holes near expended test locations indicate some local hydrogeologic effects from underground nuclear detonations at the NTS. Staged drilling and straddle-packer testing confirm an upward hydraulic gradient in the tuff confining unit. Because of the complicated hydrogeologic setting, determination of the water table has been difficult. Studies conducted to date suggest that movement of radioactive contamination over distances of kilometers is relatively rare and probably is confined to situations where there is substantial groundwater flow. After 804 underground nuclear device detonations at the NTS, no contamination of the regional aquifer has been found in any of the more than 50 wells monitoring the groundwater leaving the NTS.

This work was supported by the United States Department of Energy.

## 6. References


1. Orkild, P. P., 1983, Geology of the Nevada Test Site, For: B. C. Hudson, E. M. Jones, C. E. Keller, and C. W. Smith, (compilers), Proceedings: Monterey Containment Symposium, Los Alamos National Laboratory Report LA-9211-C, v 1: 323-338.
2. Frizzel, V. A. Jr., and Shulters, J., 1990, Geologic Map of the Nevada Test Site, Southern Nevada, U. S. Geological Survey, Miscellaneous Investigations Map I-2046.
3. Carr, W. J., 1990, Styles of Extension in the Nevada Test Site Region, Southern Walker Lane Belt; An Integration of Volcano-tectonic and Detachment Faults, The Geological Society of America, Memoir 176, Chapter 13: 283-303.
4. Winograd, I. J., and Thordardson, W., 1975, Hydrogeologic and Hydrochemical Framework South-central Great Basin, Nevada-California, with Special Reference to the Nevada Test Site: U. S. Geological Survey Professional Paper 712-C, 126 pp.
5. Arteaga, F. E., Savard, C. S., Johnson, M. E., Stone, J. C., 1991, Hydrogeologic Data from Selected Wells and Test Holes In and Adjacent to the Nevada Test Site, Nye County, Nevada Through 1986, U. S. Geological Survey Open-File Report 87-536.
6. Nimz, G.J. and Thompson, J.L., "Underground Radionuclide Migration At The Nevada Test Site," U.S. Department Of Energy Document DOE/NV-346 (1992)
7. Bryant, E.A., "The Cambric Migration Experiment," Los Alamos National Laboratory Report LA-12335-MS (September 1992)
8. Sawyer, D.A., Thompson, J.L., Finnegan, D.L., "The Cheshire Experiment, A Summary Report," Los Alamos National Laboratory Report LA-13555-MS (In Press)
9. Kersting, A.B., Efurd, D. W., Finnegan, D. L., Rokop, D. J., Smith, D. K., Thompson, J. L., "Migration of Plutonium in groundwater at the Nevada Test Site," Nature Vol. 397, pp. 56-59 (1999).
10. Smith, D.K., "A Recent Drilling Program To Investigate Radionuclide Migration At The Nevada Test Site," J. Radioanal. Nucl. Chem. 235 159 (1998)
11. Garber, M. S., 1963, Large rise and following decline of water level in Well 7, Area 3, Nevada Test Site--An effect from the Aardvark underground nuclear explosion: U.S. Geol. Survey Tech. Letter Yucca-48, 9 pp.
12. Hale, W. E., Winograd, I. J., and Garber, M. S., 1963, Preliminary appraisal of close-in aquifer response to the Bilby event, Yucca Flat, Nevada: U.S. Geol. Survey Tech. Letter NTS-63, 32 pp.
13. Doty, G. C., and Thordardson, W., 1983, Water table in rocks of Cenozoic and Paleozoic age, 1980, Yucca Flat, Nevada Test Site, Nevada, U.S. Geol. Survey, Water Resources Investigations Report 83-4067, map.
14. Thompson, J. L., 1991, Interim Report of the Drilling History and Analysis for Nevada Test Site Hole UE3e#4, Los Alamos National Laboratory Report LA-12042-MS, 203 pp.
15. Hawkins, W. L., Cavazos, A. P., and Thompson, P. H., 1987, Geologic and hydrologic investigations at the ALEMAN (U3kz) site, and other sites in Yucca Flat, the Nevada Test Site, Proceedings: Fourth Symposium on Containment of Underground Nuclear Explosions, Lawrence Livermore National Laboratory CONF-870961, v. 2: 18-26.

16. Conaway, J. G., 1987, Temperature logging as an aid to understanding groundwater flow in boreholes: Proceedings: Second International Symposium of Borehole Geophysics for Minerals, Geotechnical, and Groundwater Applications, Minerals and Geotechnical Logging Society, p 51-59.
17. Thordarson, W., and Winograd, I. J., 1962, Drill-Stem Tests of Strata Penetrated in Drill Hole Ue4a, Nevada Test Site, U. S. Geological Survey TECHNICAL

# CHARACTERISATION OF RADIONUCLIDE CONTAMINATION AT THE SEMIPALATINSK NUCLEAR TEST SITE TERRITORY

K.K. KADYRZHANOV, S. KHAZHEKBER, I.V. KAZACHEVSKY,
S.N. LUKASHENKO, V.P. SOLODUKHIN
*Institute of Nuclear Physics,*
*National Nuclear Centre of the Republic of Kazakhstan,*
*Almaty, 480082, Kazakhstan*

## Abstract

This report presents the results of work carried out at the Opytnoe Pole, the mountain massif Degelen location of the Semipalatinsk Nuclear Test Site. This work has resulted in conclusions about the characteristics of the spreading of radionuclides from air and ground nuclear explosions. Conclusions have also been reached about the level of contamination of nearby areas of tunnels, about the localisation and migration capabilities of radionuclides, and about the fractions of distribution of these radionuclides in soil.

## 1. Introduction

After nuclear testing was stopped in the territory of the Semipalatinsk test site (SNTS), numerous investigations were carried out. At present, the results of these studies of the radioecological conditions at the test site are known [1-7]. However, there was no goal-directed approach to investigating the character of the radionuclide contamination of the vast territory of the SNTS (18 500 km$^2$) resulting from these types of nuclear tests. At the same time, it is possible to characterise every kind of test by the amount and composition of the radionuclides and by the dimension, structure, and the specific activity of particles formed during the process of nuclear explosion. The area of radionuclide spreading and the localisation of radionuclides at different plots are also indicators of the type of test.

The transfer by air flow of the products of air and ground nuclear explosions and their deposition on the ground have caused contamination by sources of ionising radiation as a result of the radioactive trace in broad regions that extend for hundreds of kilometres [1].

At the SNTS territory, a programme of hydronuclear experiments (specific experiments with the ejection of large amount of alpha-active radionuclides) was carried out. These experiments also included an investigation of the fallout of alpha-active radionuclides, composed of a nuclear charge, onto the ground surface [8].



*S.S. Hecker et al. (eds.), Nuclear Physical Methods in Radioecological Investigations of Nuclear Test Sites*, 75-87.
© 2000 *Kluwer Academic Publishers. Printed in the Netherlands.*

Besides ground explosions, some excavational explosions with ground ejection also had an influence on the radiation situation in the SNTS territory. Such tests have also led to intensive local contamination [9].

The other local area of contamination is the mountain massif, Degelen. Although the main mass of nuclear explosion products is concentrated in this area in boiler spaces of tunnels, radionuclides are carried out onto the daylight surface together with waters coming through the cracks into either boiler spaces or into the tunnel. These waters are the main migrational agents and carriers of radionuclides from the sites of underground nuclear explosions [10].

This paper presents the results of work carried out within the framework of the Project K-053-96 of the International Science and Technology Centre (ISTC). This work focused on the radionuclide spreading on territories, radionuclide composition, and the concentration and localisation of radionuclides in different areas. The distribution of radionuclides in soil fractions and the migration capability of radionuclides are also discussed.

## 2. Programme of Experiments

For the study of the content and character of radionuclides distribution in soils and their fractions from ground and air nuclear explosions, eight radial direction from the centre of Opytnoe Pole were chosen.

Figure 1 presents the position of the test location as well as the traces and schemes of the radioactive fallout axes, which determined the probe selection for laboratory analysis. The heavy line shows the axes of the explosion traces, which made their main contribution to the contamination of the SNTS territory.

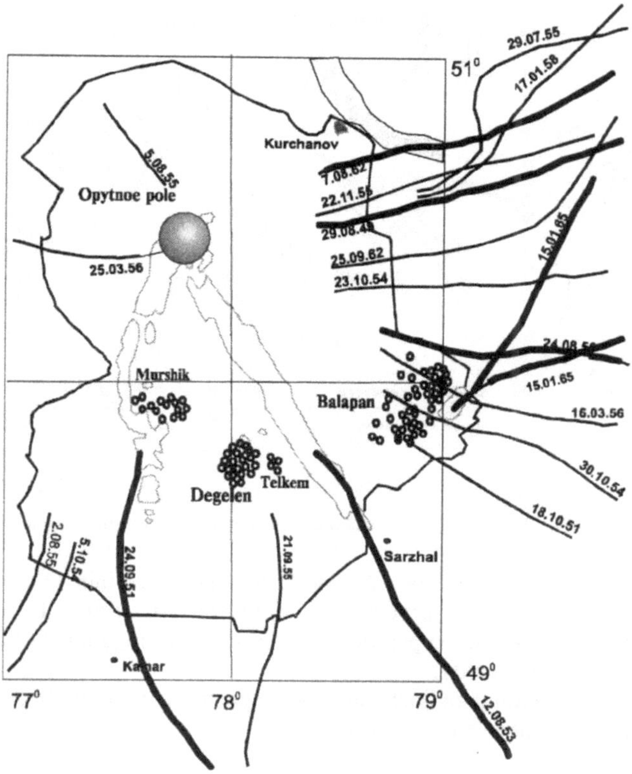


*Figure 1*. Scheme of most of the brooks at the Degelen mountain massif.

In the southeastern (SE) direction, an investigation was conducted along the trace formed as a consequence of the test of the first thermonuclear charge, with a power of 400 kt, carried out on August 12,1953. The southwestern (SW) investigation was taken along the trace formed as a result of the test of the plutonium charge, with a power of 38 kt, carried out on September 24,1951. The eastern (E) direction was taken mainly along the trace formed from the first test of the plutonium charge, with a power of 22 kt, carried out on August 29,1949. The northeastern (NE) direction was taken along the trace formed from the test of the charge, with a power of 10 kt, carried out on August 7,1962. The northwestern (NW) direction was taken along the trace formed as a result of the test of the charge, with a power of 1.2 kt, carried out on August 5,1955. Along these directions one can expect to find the spread of radionuclides from these hydronuclear experiments.

At the mountain massif Degelen, the probes were selected at the mouth of nearby tunnels with water appearing along the beds of brooks running out from the tunnels (Fig.2) along an axis, perpendicular to the first axis, with maximum contamination.

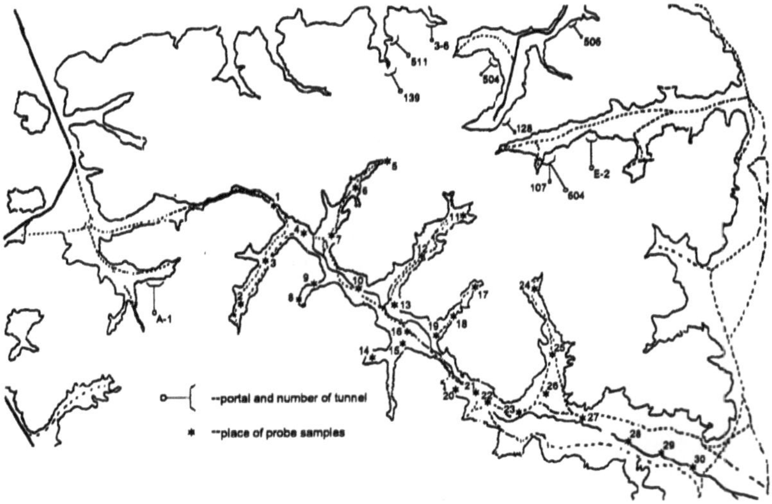


*Figure 2.* Scheme of the location of significant traces of contamination from atmospheric explosions.

During the probe selection for laboratory analysis, a certain level of radioactive contamination was taken into account. Measurements determine boundaries of the region where values of measured parameters exceed the average level of background by 1.5–2 times at the given locality.

During the investigations radiochemical and spectrometrical methods and apparatus mentioned in [11] and sieve analysis were used.

### 3. Results and Discussion

Determining the radionuclide composition of soil probes, selected at different directions and at different distances from the centre of the Opytnoe Pole, have made it possible to characterise the features of the distribution for either caesium or plutonium radionuclides at the different plots on the test site.

Figure 3 presents the dependence of the concentration of $^{137}Cs$ in soil along different directions and at different distances from the Opytnoe Pole. At the SW trace of the radioactive fallout, formed on September 24, 1951, from the testing of a plutonium nuclear charge at the power of 38 kt, the maximum content of $^{137}Cs$ was found to be up to 1300 Bq/kg. At the SE trace of the radioactive fallout formed on August 12, 1953, from the testing of a plutonium nuclear charge, at the power of 400 kt, the maximum content of $^{137}Cs$ at the distances of 26-64 km from Opytnoe Pole was found to be up to 570 Bq/kg. In the remaining six directions, up to 30km, and at the SE trace, at 26 km, the distribution has a common characteristic, and the concentration of $^{137}Cs$ is in the range of up to 100 Bq/kg. The character of the distribution of $^{90}Sr$ along these directions corresponds mainly to that for $^{137}Cs$ (Fig.4), and the ratio of their concentrations of $K_{Sr/Cs} \approx 0.1 \div 10$. Concentration of $^{241}Am$ is in the range of up to 200 Bq/kg. The maximum reading of $^{241}Am$ is found at the S direction and reaches 14 700 Bq/kg.

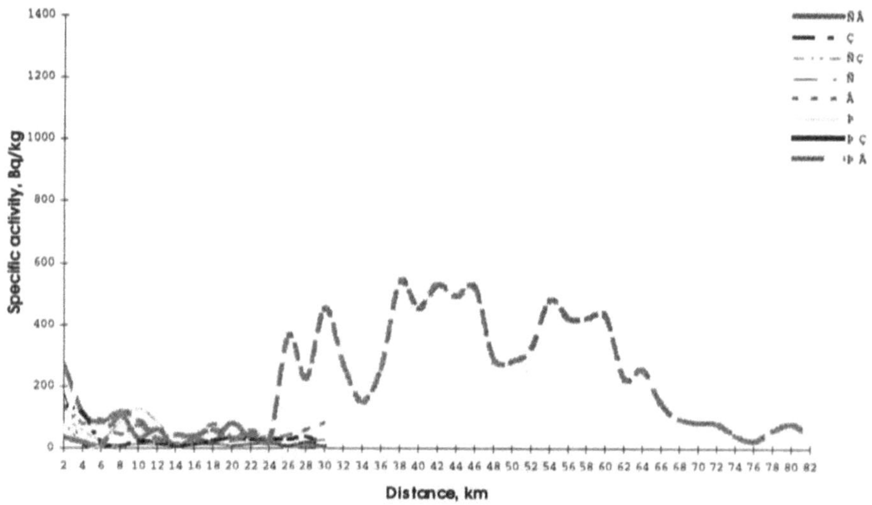


*Figure 3*. Concentration of $^{137}$Cs in soil located in different directions from the Opytnoe Pole.

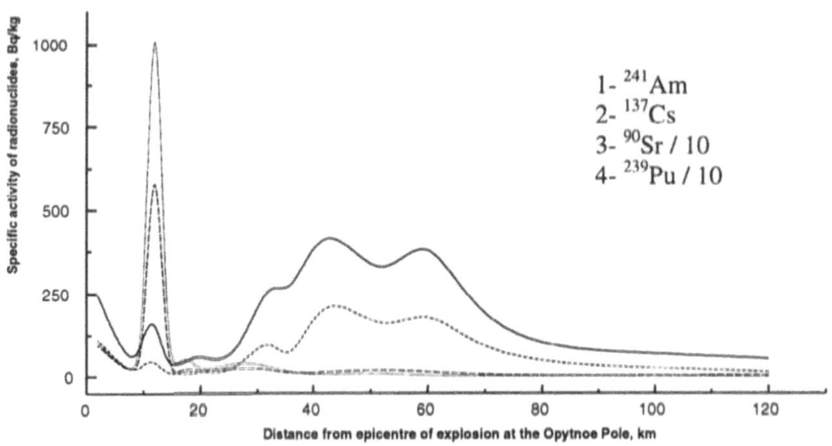


*Figure 4*. Distribution of radionuclides along the trace of SE radioactive fallout.

The character of the distribution for $^{239+240}$Pu (further $^{239}$Pu) is highly heterogeneous and differs from the distribution of other radionuclides. The northern direction is the least contaminated. Along the NE, SE, and S directions from the Opytnoe Pole, the concentration of $^{239}$Pu, at distances of 8÷20 km from the epicentre, reaches 128 000 Bq/kg (indicating a spot with large activity). Along these directions there are also spots with lesser concentration of plutonium, ranging from 1000 to 4500 Bq/kg. At the SW trace, where the explosion was carried out from a tower of 30 m high, the total ejection of plutonium was 300 Ci [12]. The character of distribution for $^{239}$Pu has several maxima (from 2000 to 7000 Bq/kg) at distances of 8, 16, 22-24, 34, and 40 km from the centre. Generalising from the above-stated facts, one can say that the contamination by plutonium has a spotted character and extends locally along the S, NE, SW, SE, E, and NW directions from the Opytnoe Pole (Fig.5).

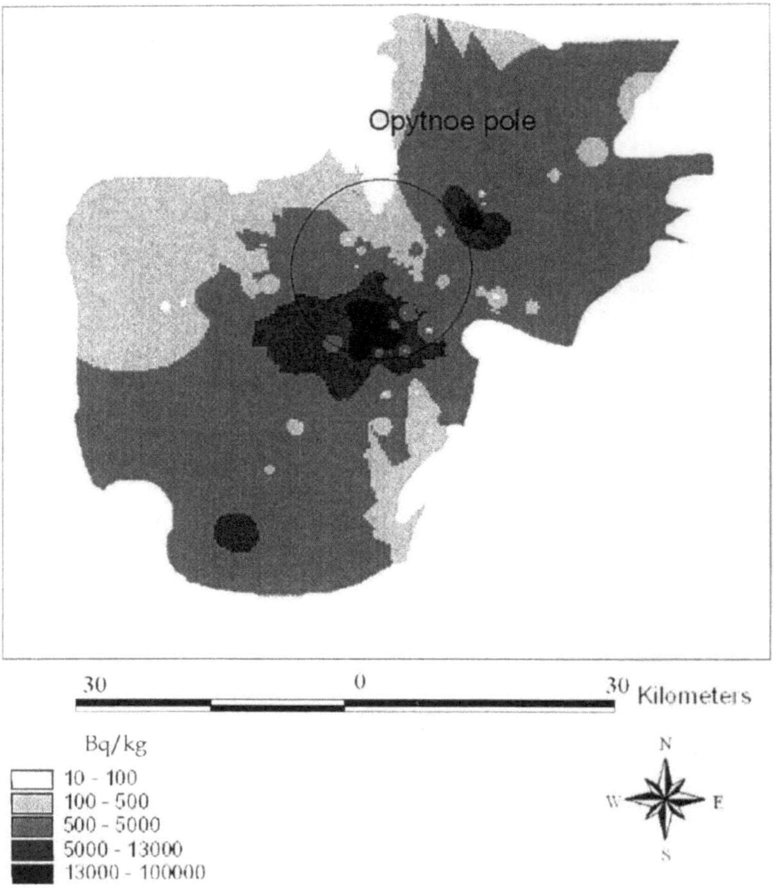


*Figure 5*. Distribution of Pu-239/240 concentrations in soil from ground explosions at the SNTS.

Naturally the significant contents of plutonium at the Opytnoe Pole and its surrounding areas attract close attention. It is necessary, too, to pay attention to the character of plutonium distribution as it differs from the distribution of other radionuclides. One of the characteristics helping one to ascertain the nature of radioactive contamination can be the value of the ratios of radionuclide concentrations. The ratios of concentrations for $^{239}$Pu and $^{137}$Cs in soil samples selected from this territory are presented in Figure 6. In all of the samples investigated, this ratio varies up to a limit of five orders. One can note that for background levels of global fallout (4-5 kBq/m$^2$ for $^{137}$Cs and 60 Bq/m$^2$ for $^{239}$Pu [13]), this value is ~0.01. The largest values of the ratio of concentrations for $^{239}Pu$ and $^{137}Cs$ correspond to the highest concentrations of $^{239}Pu$. Very probably, the areas indicated above with large plutonium deposition came to be under ejection from hydronuclear experiments. So, one can conclude that the plutonium contamination in the Opytnoe Pole area has a locally-specific character and, undoubtedly, requires further closer definition and an acceptance of some measures for remediation of these areas.

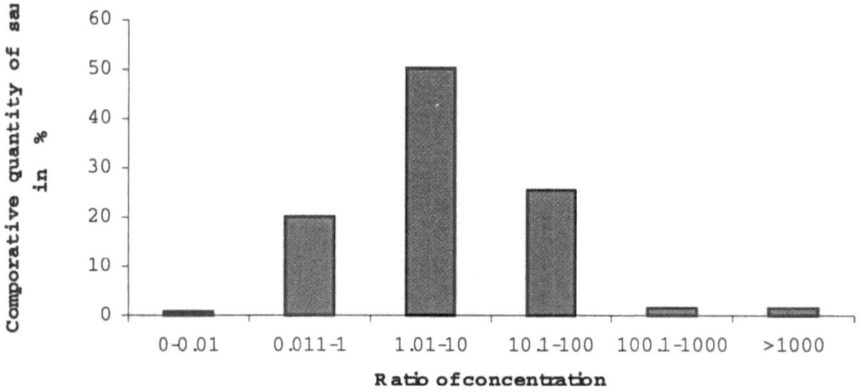


*Figure 6.* Distribution ratios of the concentration of $^{239}Pu$ to that of $^{137}Cs$.

Figure 7 presents data about the distribution of the $^{137}Cs$ concentration on nonmagnetic granulometric fractions (Bq/kg). It also shows the abscissa: a size of fractions, and ordinates: a concentration of $^{137}Cs$, (Bq/kg). Digits above the columns of data show a concentration of $^{137}Cs$(Bq/kg) for given fractions. These fractions had different masses: from hundreds of grams for clay (600-800g) to 20-50 grams (for the majority of cases) for other fractions.

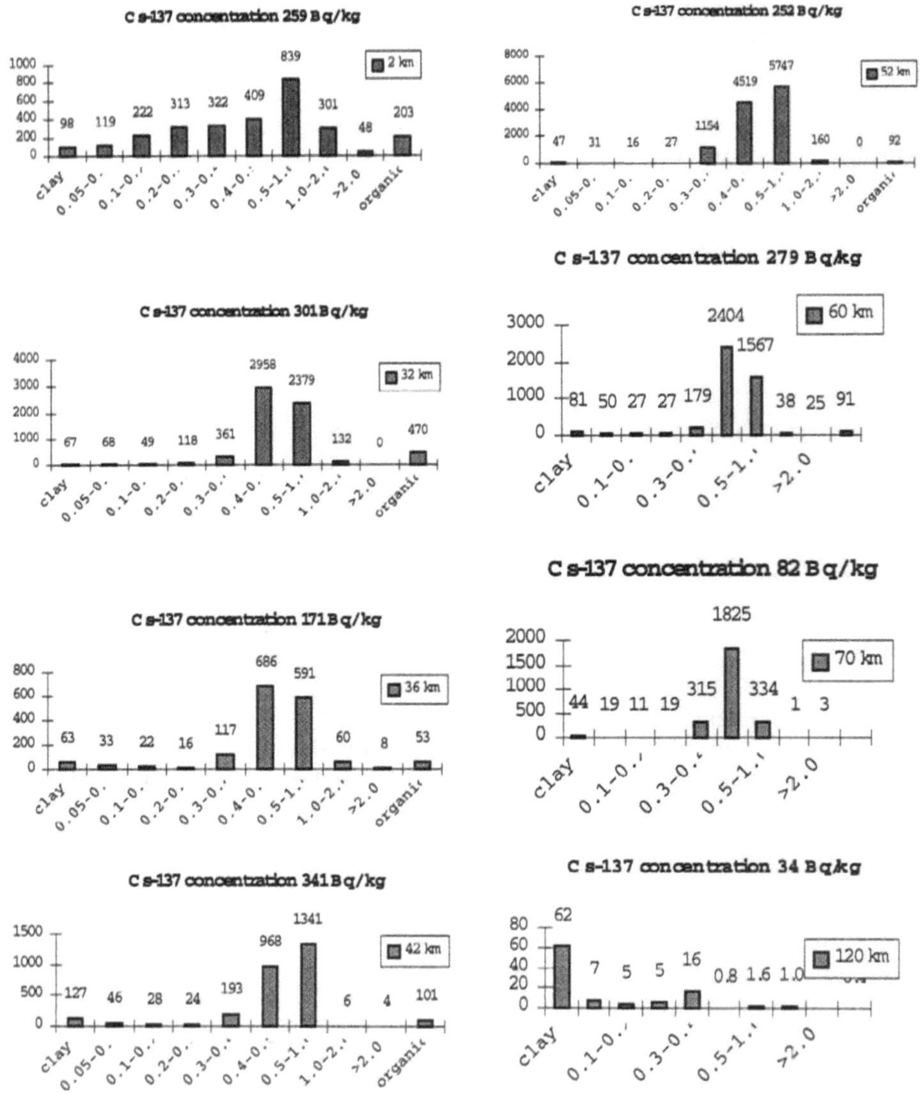


*Figure 7.* Distribution of $^{137}Cs$ concentrations in soil fractions in the SE direction (Bq/kg).

A concentration of $^{137}Cs$ varies widely and reaches the level of several thousand Bq/kg. (A large value of a concentration of $^{137}Cs$ for an "organic" fraction is connected with a small mass of a measured sample.) However, a small mass of these fractions leads to a less than large total activity of the sample. It is interesting to note that for the finest fraction (clay), the concentration changes three times for all samples—from 127 Bq/kg to 44 Bq/kg, with the average concentration being 70 Bq/kg. However, the concentration of $^{137}Cs$ for the entire sample changes by one order (Fig.8). One can suppose that the process of air transfer "averaged" the concentration of $^{137}Cs$ in the investigated territory. On the basis of this supposition, one can assume that now the investigated territory of the SNTS does not show more significant secondary contaminant because of the wind transfer. To complete a detailed study of this assumption, the relative distribution of contents of $^{137}Cs$ on fractions, i.e. the relative contribution (in %) of a given fraction to the total activity of the probe, was investigated. For the majority of cases, in all samples, irrespective of the total contents of $^{137}Cs$ in the soil, the main activity in fractions ranged from 0.3 to 1.0 mm. This finding shows impressively enough that, for the most contaminated plots along the SE trace of the SNTS, the main contributions to the total activity of samples are made by rather large particles (>0.3 mm), for which migration processes are difficult.

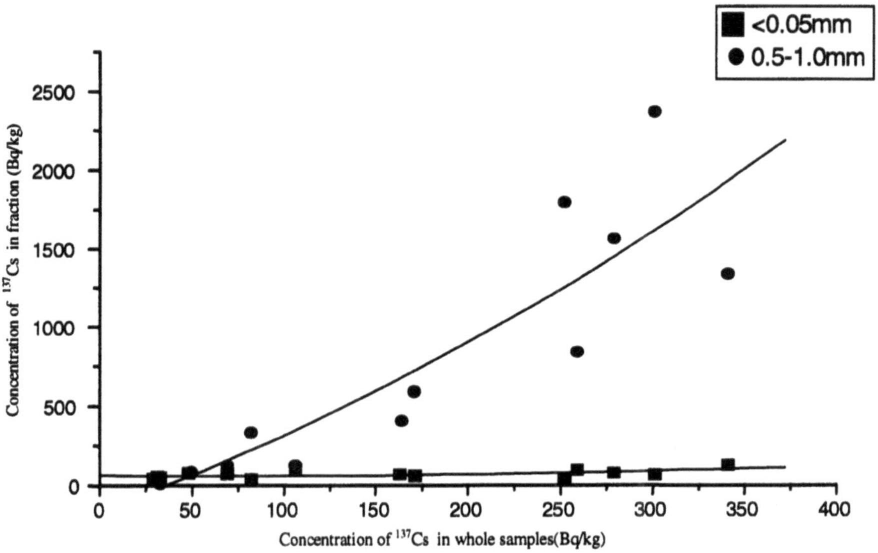


*Figure 8.* Changes in the concentration of $^{137}Cs$ in small and big fractions, irrespective of the total content in the sample.

Table 1 shows selective results of the analysis of radionuclide composition of soils selected at nearby tunnel entrances along brooks at the mountain massif Degelen and at different soil depths. From the data shown, it follows that the specific activity of radionuclides varies widely:

- $^{239}Pu$: from several tens of $n\times10^5$ Bq/kg;
- $^{90}Sr$: from $n\times10^2$ to $n\times10^5$ Bq/kg;
- $^{137}Cs$: from $n\times10^2$ to $n\times10^6$ Bq/kg;
- $^{241}Am$: from tens to $n\times10^4$ Bq/kg.

TABLE 1. Radionuclide concentrations in the soil at the near-by entrance areas of the Degelen mountain massif (Bq/kg).

| Tunnel number, probe selection place | $^{239+240}Pu$ | $^{90}Sr$ | $^{137}Cs$ | $^{241}Am$ |
|---|---|---|---|---|
| Z- 6 near-by-mouth area | $9.1\times10^2$ | $4.3\times10^4$ | $5\times10^6$ | $7.7\times10^2$ |
| | $1.1\times10^3$ | $3.7\times10^4$ | $6\times10^6$ | $2.8\times10^2$ |
| 139 near-by-mouth area | $8.6\times10^5$ | $6.9\times10^2$ | $6.2\times10^2$ | $6.6\times10^4$ |
| | $1.7\times10^4$ | $1.1\times10^2$ | $1.6\times10^2$ | $2.3\times10^3$ |
| 504 near-by-mouth area | $3.5\times10^2$ | $1.3\times10^5$ | $1.9\times10^6$ | <15 |
| | $4.3\times10^4$ | $0.9\times10^2$ | $1.7\times10^2$ | <15 |
| 506 near-by-mouth area | $5.2\times10^2$ | $1.2\times10^2$ | $0.6\times10^2$ | <15 |
| | $1.2\times10^4$ | $2.0\times10^4$ | $3.1\times10^5$ | $8.3\times10^2$ |
| -1 near-by-mouth area | $1.1\times10^4$ | $2.3\times10^5$ | $2.0\times10^5$ | $8.9\times10^2$ |
| | $7\times10^2$ | $1.1\times10^5$ | $1.0\times10^5$ | $2.3\times10^2$ |
| 511 near-by-mouth area | 34 | $5.7\times10^4$ | $2.6\times10^5$ | 70 |
| | 55 | $9.0\times10^4$ | $2.4\times10^5$ | 40 |
| 350 m from near-by-mouth | | | | |
| 0-5 m | 78 | $3.2\times10^4$ | | |
| 5-10 m | 11 | $9.2\times10^3$ | | |
| 10-15 m | 133 | $1.8\times10^4$ | | |
| 15-20 m | 16 | $7.1\times10^3$ | | |
| Along of the brook "Uzyn-Bulak" | $1.5\times10^2$ | $3.0\times10^4$ | $6.9\times10^2$ | <15 |
| | $9.5\times10^2$ | $7.0\times10^4$ | $3.1\times10^5$ | <15 |
| | $6.8\times10^4$ | $5.3\times10^2$ | $2.7\times10^4$ | 35 |
| | $5.5\times10^4$ | $1.8\times10^3$ | $1.3\times10^3$ | $3.3\times10^2$ |

For radioactive contamination at areas around all of the tunnels, the different concentrations of radionuclides are inherent.

Figure 9 presents the values of the ratios of specific activities of radionuclides ($^{137}Cs/^{90}Sr$ and $^{137}Cs/^{239}Pu$) in soil samples that were selected at different tunnels and subjected to radiochemical analysis. It is known that during the fission of $^{235}U$ and $^{239}Pu$ nuclei, the ratios of the activities of $^{137}Cs/^{90}Sr$ are in the interval (1.1-3.2).

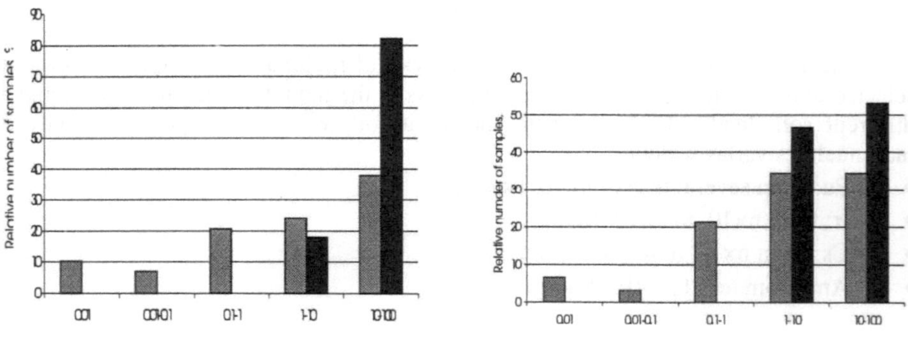


*Figure 9.* Ratios of the radionuclide concentrations in soil probes from near-by tunnel entrance areas (the tunnel with water).

From the information in Figure 9 it follows that for the majority of tunnels, this ratio is essentially different. This variation can be caused by a number of factors, such as: features of nuclear charge; an irregular (accidental) situation, accompanied by the escape of explosion products onto daylight surfaces; an amount of underground water running onto daylight surfaces through the nearby entrance of the tunnels; sorptional and desorptional capacities of soil; migration capacities of the radionuclides.

The results (obtained with indicators: $^{137}Cs$ and $^{85}Sr$) of sorptional and desorptional capacities of soil samples, selected at nearby entrances to tunnels 501 and 511 (see Table 2), show that they possess a significant sorptional capacity for Cs and a lesser capacity for Sr. The desorption of Cs is small; at the same time, however, $^{85}Sr$ is desorbed to a considerable extent.

TABLE 2. Sorptional and desorptional capacities of soil

| Tunnel | Type of soil | Sorption, % | | Desorption, % | |
|---|---|---|---|---|---|
| | | $^{137}Cs$ | $^{90}Sr$ | $^{137}Cs$ | $^{90}Sr$ |
| 501 with water manifestation | loam | 94.1 | 22.6 | 3.7 | 66.7 |
| 511 with water manifestation | sod | 98.7 | 53.5 | 1.6 | 32.8 |

From Figure 9 one can note that at the nearby entrance areas of tunnels with stable water manifestation $^{137}Cs$ is concentrated more than are $^{239}Pu$ and $^{90}Sr$.

Figure 10 illustrates a change of radionuclide contents at different distances from the nearby entrance of tunnel 609 along the bed of the brook. The figure shows that directly at the nearby entrance, the levels of concentrations of plutonium, strontium, and caesium are comparable: ~$10^5$ Bq/kg. But the farther one moves from the entrance, the concentration of plutonium is reduced by a factor of $10^3$. Relatively high contamination by plutonium, caesium, and strontium was also discovered in places sufficiently removed from the nearby entrance (up to 2.5 km and farther) of the tunnel. This fact can be explained by assuming that essential migration of these radionuclides occurs. To all appearances during the process of migration, there is an essential redistribution of radionuclides that is connected with the relation between acid-soluble and acid-insoluble forms of radionuclides in soil probes.

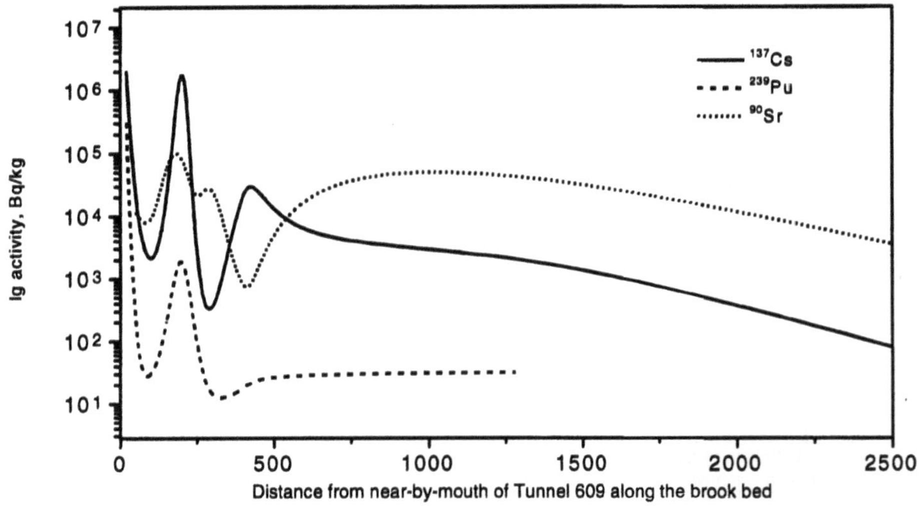


*Figure 10.* Change of radionuclide concentrations along the brook bed.

On the basis of the comparison of changes of isotope ratios in points located at different distances from the same source of contamination, one can make a conclusion about the comparative migrational capacities of radionuclides. This finding shows that the investigated radionuclides can be put in the following order: strontium > caesium > plutonium.

Estimating the results of laboratory investigations and field work performed at the mountain massif Degelen, one can make some preliminary conclusions. Tunnels containing water are the main sources of contamination of the adjoining territories. The contamination of nearby entrance areas and stream beds by radionuclides is very high. (At some separate points, concentrations measured in Bq/kg reach: 860 000 for plutonium, 230 000 for strontium, and 6 000 000 for caesium.) Radionuclides (mainly, $^{90}$Sr) are water-soluble and leach out from underground spaces in the form of ions that define their high-migration capacities. (Even when removed from the nearby entrance up to 2.5 km and farther, concentrations of plutonium, strontium, and caesium remained high.)

## 4. Conclusions

As a result of the investigations that have been carried out in this region, radionuclide distribution from ground and air nuclear explosions has migrated in different directions from the centre of the Opytnoe Pole. The locally specific character of the distributions of weapon plutonium has been determined. Also established are the contamination levels at the nearby entrance areas of tunnels and of stream beds at the mountain massif, Degelen. The composition and migration capacities of radionuclides that lead to radionuclide transfer over considerable distances are also known.

Data about distribution of radionuclides on granulometrical soil fractions have been obtained as were the important results for further developing and choosing the technology for removing surface contamination of these territories.

The work described in this paper is being carried out within the framework of Project K-053-96, at the International Scientific and Technology Centre (ISTC), with the collaboration of the Los Alamos National Laboratory (LANL), USA.

## 5. References

1. Report about carrying out of airgeophysical work complex in the region of the SNTS (1990) MKAE PGO "Aerogeologia", (in Russian).
2. Dubasov, Yu.V. (1997) Contemporary radiation situation at the former SNTS and round it. I. Some results of control-methodical inspection in 1994. Radiochemistry, **39**, 80-88, (in Russian).
3. Tipton, W.J. (1994) Kazakhstan in situ Radiation Survey, Rep.RSL-9401
4. Burmistrov, V.R., Kazachevskiy, I.V., Tukhvatulin, Sh.T. etc. (1996) Field Measurements, Soil Sampling and Laboratory Analysis for Ten Areas on the Former Semipalatinsk Nuclear Test in the Republic of Kazakhstan, Almaty.
5. Preliminary estimation of radiological situation at the STS, Republic of Kazakhstan (1996) Report, IAEA, Vienna.
6. Ptitskaya, L.D. etc. (1996) The study of radiation situation at the territory of the former SNTS, Report of INSE, Kurchatov, (in Russian).
7. Yamamoto, M. etc. (1996) Preliminary Results of the Present Situation of Radioactive Fallout in Soil at SNTS. Proc. of the Second Hiroshima Intern. Sympos, 201-224, Health Physics **71**, 142-148.
8. Andrushina, I.A. etc. (1998) Nuclear Tests of USSR. Hydronuclear experiments. Inventory of plutonium expenditure, RPhYaTs-VNIIEPh, PhTTs, Sarov, (in Russian).
9. Akhmetov, .M. etc. (1994) A characteristic on initial data of radiation situation of epicentral zone of the object «Chagan» – of excavational underground nuclear explosion for creation of artificial water storage, Izvestiya of NAN RK, 79-96, (in Russian).
10. Smagulov, S.G. (1992) Information on results of radiation inspection of the territory of the mountain massif Degelen, connected with presence of water manifestation in tunnels, (in Russian).
11. Kazachevskiy, I.V., Lukashenko, S.N. etc. (1999) Combined radiochemical procedure for determination of plutonium, americium and strontium-90 in the soil samples from SNTS, Czechoslovak Journal of Physics, **49**, Suppl. S1.
12. Dubasov, Yu.V. etc. (1994) Chronology of nuclear test in atmosphere at the SNTS and their radiation characteristics, "The Bulletin of the scientific programme STS Altai", pp. 77-86, (in Russian).

# RADIOACTIVE STRONTIUM- AND CAESIUM-CONTAMINATED SITES: CHARACTERISATION, TRANSPORT, AND REMEDIAL OPTIONS

C.F.V. MASON
*Nuclear Materials Technology Division*
*Los Alamos National Laboratory, Los Alamos, New Mexico 87545, U.S.A.*

N. LU AND J. CONCA
*Environmental Technology Division*
*Los Alamos National Laboratory, Los Alamos, New Mexico 87545, U.S.A.*

## Abstract

Radioactive strontium (Sr) and radioactive caesium (Cs) are contaminants causing major concern because of their ease of entry into biological systems due to their similarity to calcium and potassium, respectively. As a result of the Chernobyl accident in 1986, large areas of the Ukraine are contaminated with these radionuclides. Areas both at the Los Alamos National Laboratory (LANL) in the United States and at the Semipalatinsk Test Site in Kazakhstan are also contaminated with these radionuclides. Both caesium and strontium are soluble in water, but their mobility in natural environments is retarded by their strong interactions with soil organic materials, clay minerals, and iron oxides.

In this paper we examine a site at LANL contaminated with $^{137}Cs$ and $^{90}Sr$. Experimental results show that the distribution coefficients for Cs are much greater than those for Sr. This difference results in much faster dispersal of Sr through water-borne transport pathways. Furthermore, Sr is transported by sorption onto colloids, thereby providing an additional transport mechanism.

Remedial options for contaminated sites require detailed studies of sorption processes of the local soils. Apart from the drastic measure of removing contaminated soils to a controlled area, the movement of contaminants can be controlled by the presence of media that will prevent both particulate movement and also immobilise soluble species. Immobilisation occurs through precipitation, sorption, and ion exchangers.



*S.S. Hecker et al. (eds.), Nuclear Physical Methods in Radioecological Investigations of Nuclear Test Sites*, 89-97.
© 2000 *Kluwer Academic Publishers. Printed in the Netherlands.*

## 1. Introduction

Radioactive strontium (Sr) and radioactive caesium (Cs) enter the environment through above-ground nuclear weapons tests, nuclear accidents, and the improper disposal of waste from the processing of nuclear fuel rods. These fission products cause great concern because of their easy uptake into biological organisms due to their similarity to calcium and potassium, respectively. Caesium-137's half-life is 30 years and Sr–90's is 28.5 years. The migration of these radionuclides in soils is an important consideration since soils have a close relationship to human life. In addition, together these radionuclides are responsible for 98% of the thermal energy and 97% of the penetrating radiation in high-level radioactive waste.

As a result of the Chernobyl accident in 1986 [1], large areas of the Ukraine are contaminated with these radionuclides. Of the total release of ~50 MCi, 70% of the activity fell on 17 000 $km^2$ of a forested area of Gomel Oblast in Belarus. In May 1996, it was estimated that 795 200 Ci of $^{137}Cs$ and 172 900 Ci of $^{90}Sr$ remained.

Areas both at the Los Alamos National Laboratory (LANL) [2] in the United States and at the Semipalatinsk Test Site in Kazakhstan [3] are also contaminated with these radionuclides. Strontium–90 is the third most frequently occurring radionuclide in ground water at the US Department of Energy facilities [4]. Furthermore, the sorption of radionuclides is an important parameter in the design of nuclear waste repositories.

In this paper we discuss a case study in the characterisation and remediation of a Cs- and Sr-contaminated site. In addition we describe further remedial options.

## 2. Description and Characterisation of the LANL Site

At Los Alamos, New Mexico, DP Canyon was subject to radioactive contaminant releases from 1945 to 1978 from the Plutonium Processing Facility that was used for research and the production of plutonium [5]. Radioactive effluent, which was discharged down the steep slope of the adjacent canyon to the north, DP Canyon, resulted in extensive contamination. Surveys of this area (including an aerial radiation survey) identified a plume of radionuclides including $^{137}Cs$, $^{241}Am$, $^{239}Pu$, and $^{90}Sr$. These are stored in the thin soils located down the side of the canyon and also in the sediments of the stream channel at the canyon bottom. The highest contamination levels, (found to be due to $^{137}Cs$) were located on the steep slope of the canyon wall directly below the discharge point [5]. This configuration is illustrated in Figure 1. This site is named PRS 21-011(k).

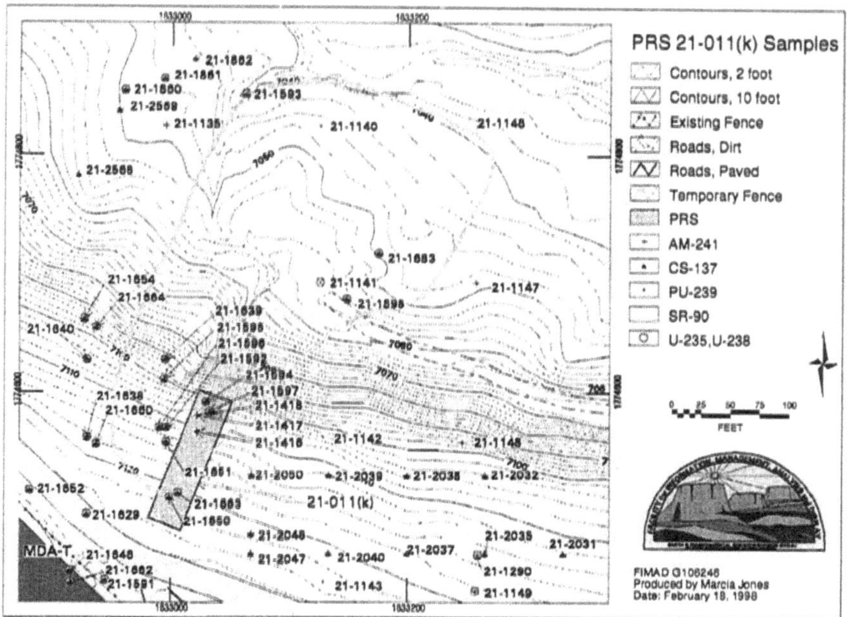


*Figure 1.* Map showing the location of the PRS 21-011(k) site. Sample points and contours are shown, together with the temporary fence enclosing the area where the dose was calculated to be greater than 15 millirem/y.

## 3. Transport Mechanisms

Though both caesium and strontium are soluble in water, their mobility in natural environments is retarded by their strong interactions with soil organic materials, iron oxides, and clay minerals. In general the sorption distribution coefficients ($Kd_S$, ml $g^{-1}$) are calculated according to the following equation:

$$Kd_S = [(A_0V_0 - A_fV_f)/M]/A_f \qquad (1)$$

where $A_0$ is the initial radioactivity of $^{137}$Cs or $^{90}$Sr per millilitre solution,
$V_0$ is the initial volume (ml) of $^{137}$Cs or $^{90}$Sr solution in contact with solid samples,
$A_f$ is the final radioactivity per millilitre solution after sorption,
$V_f$ is the final volume (mℓ) of the solution after sorption, and
M (g) is the mass of solids or colloids used in the sorption process.

As shown in Table 1, experimental results from the soils and sediments in Los Alamos County [6] reveal that the distribution coefficients for Cs are much greater than those for Sr. Consequently, dispersal of Sr through water-borne transport pathways is much faster, while Cs is mainly transported with the sediments and soils.

TABLE 1. Distribution Coefficients for Cs and Sr [6]

| | Cs | Sr |
|---|---|---|
| Surface soils | $Kd_s$ = 160 -1400 ml/g | $Kd_s$ = 15 -70 ml/g |
| Sediments | $Kd_s$ = 80 -1000 ml/g | $Kd_s$ = 10 -40 ml/g |
| Desorption | 4-9 % | 8 - 30% |

It was found that the radioactivity associated with Cs decreased linearly as the soil's particle size increased, suggesting that Cs is mainly associated with the silt-clay fractions [5]. The silt and clay fractions of Los Alamos soils are largely montmorillonite and illite formed by weathering of tuff (the volcanic base for Los Alamos County). Monovalent cations such as Cs enter the interlayer of illite preferentially and become irreversibly bound. Thus, as was experimentally verified [5], Cs bound to these clay minerals is not leached by rainfall.

There is also evidence that Cs is transported by colloids [7]. Snowmelt from the base of DP canyon where the effluent was discharged was filtred through successively smaller filtres (0.45-μm, 0.2-μm, 0.05-μm nylon membranes) that removed all the radioactivity (see Fig. 2). The results show that most particles ranged from 0.05 to 0.45 μm. The dry weights of the particulates collected are 632.7 mg $L^{-1}$ with no filtre, 554.3 mg $L^{-1}$ with a 0.45-μm filtre, 521.5 mg $L^{-1}$ with a 0.2-μm filtre, and 507.2 mg $L^{-1}$ with a 0.05-μm filtre. These findings indicate that the $^{137}$Cs is associated with the fine particles including colloids.

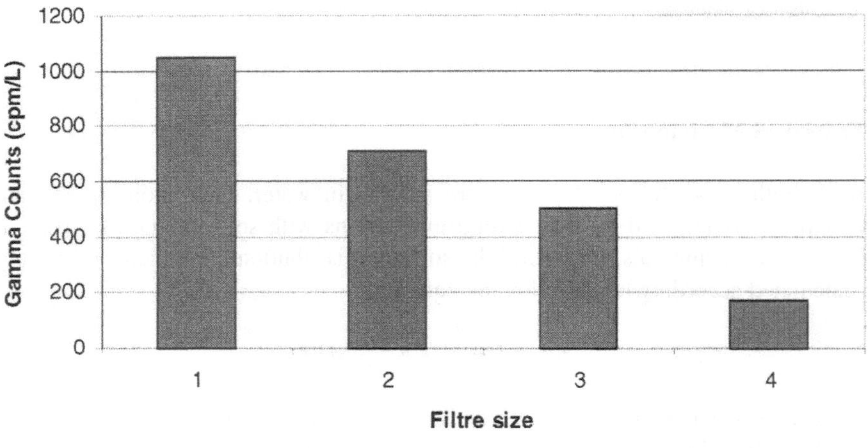


*Figure 2*. Radioactivity of snow melt water as a function of size of filtre. (1= no filtre, 2= 0.45μm, 3= 0.2 μm, 4= 0.2μm)

Conversely, $^{90}$Sr desorbs from soils and sediments continuously and has a fast migration rate relative to that of other contaminants [8]. A radiological survey in 1995 found that as much as 526 pCi $L^{-1}$ of $^{90}$Sr exist in shallow alluvial wells in DP Canyon. Furthermore, Sr is transported by sorption onto colloids, thereby providing an additional transport mechanism. However the presence of soluble cations, such as sodium and calcium, will affect the sorption properties. Figure 3 shows the sorption of $^{85}$Sr (used as a laboratory surrogate for $^{90}$Sr) onto two colloids using deionized water and natural water from DP Canyon. Calcium in DP canyon water competes successfully for the sorption sites when the silica colloid is used but is unaffected for the smectite (calcium montmorillonite) colloid.

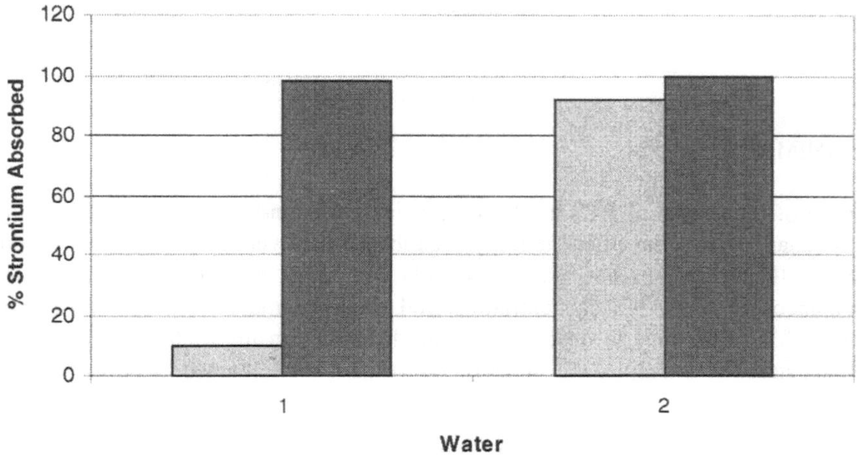


*Figure 3.* Percentage of $^{85}$Sr sorbed onto colloids of silica-ST-ZL (gray) and smectitie (black) after 7days. (I indicates DP canyon water; 2 indicates deionized water)

## 4. Entry into Biological Systems

### 4.1. PLANTS

A common shrub that grows in the semiarid climate of New Mexico is the chamisa (*chrysothamnus nauseosus*) with roots growing to depths of 200 cm. Table 2 shows the uptake of $^{90}$Sr into a chamisa bush situated over a formerly low-level waste disposal area. Clearly, not only does chamisa sorb Sr, it also redistributes it through plant detritus onto the ground surface.

TABLE 2. Strontium-90 in chamisa in a former liquid waste disposal area [9]

| Uptake medium | Type of growth | Sample 1 (pCi/g dry ash) | Sample 2 |
|---|---|---|---|
| Chamisa | Top growth | 90 500 | 5 930.0 |
| | Root growth | 40 600 | 4 630.0 |
| Background | Plant top growth | 0.3 | |
| | Root growth | 0.0 | |
| Soil | Understory | 191.4 | 37.5 |
| | Interspace | 12.6 | 16.0 |
| Background | Understory | 0.1 | |
| | Interspace | 1.40 | |

4.2. ANIMALS

A study of the uptake of $^{137}Cs$ by the wild rodents was undertaken in 1978 [10] in the stream channel from the outfall that received the effluent waste in DP Canyon. Lithium fluoride thermoluminescent dosimeters (TLDs) were implanted in the animals subcutaneously. The $^{137}Cs$ concentrations and absorbed radiation doses are shown in Table 3. The difference in dose reflects the difference in the mobility and habitat of each species. For example, the Western Harvest mouse remained close to the stream channel.

TABLE 3. Mean $^{137}Cs$ concentrations and mean estimated radiation dose [10]

| Species | Number of mammals | (pCi/g wet carcass) ($^{137}Cs$) | Total dose* (mrads/day) |
|---|---|---|---|
| Western Harvest mouse | 8 | 41 | 26 |
| Deer mouse | 4 | 6.6 | 8.3 |
| Piton mouse | 5 | 1.0 | 1.9 |
| Chipmunk | 13 | 1.6 | 1.4 |

*Measured subcutaneously with an implanted TLD.

## 5. Remedial Actions

Because DP Canyon is a favourite place for outdoor activities, the first priority was to render it safe for hikers and other recreational users. This objective was accomplished in the summer of 1996 when approximately 300 $m^3$ of $^{137}Cs$-contaminated soil with $\gamma$ activity >100 000 cpm were removed, leaving the remaining activity at the site at <100 000 cpm (~ <400pCi/g). This removal included the hot area shown in Figure 1. The soil was trucked to a low-level radioactive repository at Los Alamos.

To stop the spread of $^{137}Cs$-contaminated soil, the additional remedial action was taken of placing straw bales over the area where the soil washes down the canyon wall during the short period when intense thunderstorms occur. These straw bales prevent the movement of particulate matter. As a result of this action, the average $\gamma$ activity was reduced by 59%, showing that the straw bales blocked the coarse soil particles but not the fine ones [7].

## 6. Possible Remedial Options

Caesium-137-contaminated soil particles of <2.00 mm have been shown to be blocked by a 5-layered sand barrier [5]. No detectable $^{137}Cs$ was found either below the second layer of the barrier or in the leached water. Movement of soluble contaminants, such as $^{90}Sr$, can be controlled by the presence of media that will immobilise $^{90}Sr$. Such media are shown in Table 4.

TABLE 4. Immobilisation mechanisms

| Immobilisation mechanism | Example |
|---|---|
| Precipitation | 1. Phosphates such as apatites (e.g. $Sr5\ (PO4)_3\ (OH)$, log Ks = -51.3) [11]<br>2. Organic phosphates (phytic acid) [12]<br>3. Barium sulfate (coprecipitation) [13] |
| Ion exchange and sorption | Clays, zeolites, geothite, beidellite, clinoptilolite [14] silico-titanates [15] |

### 6.1. RICHARDS BARRIERS

A Richards Barrier [16] acts in an unsaturated hydrogeologic system to prevent the advective flow of water down through a barrier. This barrier incorporates two or more layers with contrasting capillary suction potentials and hydraulic conductivities such as a fine layer of sand with small pores and high capillary potential and also a coarse layer of gravel with large pores and low capillary potential. The hydraulic conductivity of the sand causes the water to drain down the layer boundary. This process is illustrated in Figure 4. Such barriers have been found to be effective over thousands of years at prehistoric burial sites in Japan [17].

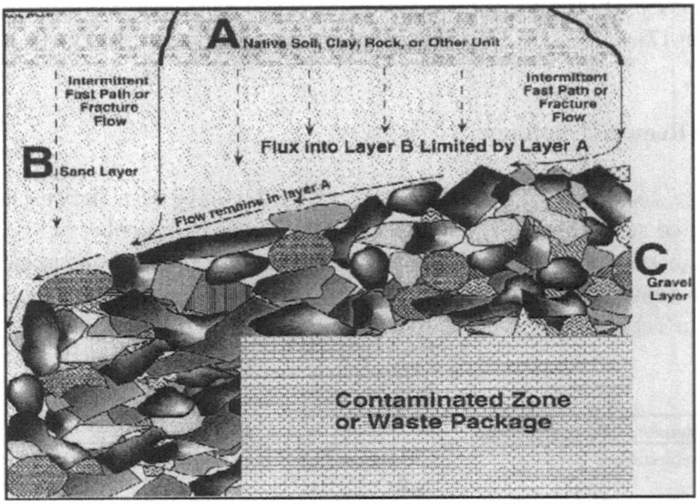


*Figure 4*. Features of a Richards Barrier (gravel barrier) for hydrologic isolation in the vadose zone [16].

## 7. Summary

By detailed examination of one site contaminated with $^{137}Cs$ and $^{90}Sr$, it is possible to reach some general conclusions as follow:

- Even when $^{137}Cs$ and $^{90}Sr$ originate from the same source, they will be unevenly distributed due to different transport mechanisms. This pattern is found at all contaminated sites, including those in the Ukraine.
- $^{137}Cs$ and $^{90}Sr$ are both taken up in biological systems.
- Both $^{137}Cs$ and $^{90}Sr$ can be controlled by barriers and immobilisation.
- Each contaminated site is unique.

## 8. Acknowledgements

The authors wish to acknowledge Nancy Schulte of NATO for sponsoring the workshop where this paper was first presented and also their colleagues at Los Alamos National Laboratory, including Danny Katzman, Steven Kung, Pat Longmire, and Ines Triay.

# 9. References


1. D. McCarn, L. Dubovik, A. Ialoushev and A. Grebenkov, "The Ecological-Commerce (Eco-Com) zone concept for developing biomass energy from contaminated resources: A new demonstration zone for the Republic of Belarus," Int. Topical Meeting on Nuclear and Hazardous Waste Management, Seattle, Washington, pp. 1417-1424, August 1996
2. Los Alamos National report. Phase report 1C(TA-21), 1994, LA-UR-94-228
3. C.F.V. Mason, N. Lu, N.L. Marusak, B. Scheber, S. Chipera, D. Daukeyev and I. Khromushin, "Strontium-85 and plutonium-239 sorption by rocks from the Semipalatinsk Test Site, Kazakhstan" Proceedings of Waste Management '99
4. R.G. Riley, J.M. Zachara and F.J. Wobber, "Chemical contaminants on DOE lands and selection of contaminant mixtures for subsurface science research," DOE/ER-0547T. US DOE, Office of Energy Research, Washington, DC, 1992
5. N. Lu, C.F.V. Mason, W.R.J.R. Turney, Characterization and immobilisation of caesium-137 in soil at Los Alamos National Laboratory," 3rd International Conference on Precision Agriculture, St. Paul, MN, June 23-26, 1996
6. K.S. Kung, B.W. Li, P.A. Longmire and M.M. Fowler, "Sorption and desorption of caesium and strontium on TA-2 and TA-41 Soils and Sediments," LA-13108-MS, April 1996
7. N. Lu, C.F.V. Mason, D. Katzman, H.D. Kitten, G. McMath and P.A.Longmire, "Characterization, transport and stabilization caesium-137 in a canyon at Los Alamos National Laboratory" Accepted for publication in Soil Science of America, June 1999
8. N. Lu, I.R. Triay, C.F.V. Mason and P.A. Longmire, "Sorption behavior of Strontium-85 onto colloids of silica and smectite," Accepted for publication in Radiochimica Acta, June 1999.
9. P.R. Fresquez, T.S. Foxx and L. Naranjo, " Strontium concentrations in chamisa (chrysothamnus nauseosus) shrub plants growing in a former liquid waste disposal area in Bayo Canyon," LA 13050-MS, November 1995
10. F.R. Miera and T.E. Hakonson, "Radiation doses to rodents inhabiting a radioactive waste receiving area," Health Physics vol. 34, pp 603-609, June 1978
11. X.B. Chen, J.V. Wright, J.L. Conca and L.M. Peurrung " Effects of pH on heavy metal sorption on mineral apatite" Environmental Science and Technology vol.31, pp 624-631, 1997
12. K.L. Nash, M.P. Jensen, J.J. Hines, S.A. Friedrich, M. Redko, "Phosphate mineralization of actinides by measured addition of precipitating anions" Argonne National Laboratory, TTP: CH2-6-C3-22, 1997
13. P. Longmire, J. Boak, F. Caporuscio and A. Adams " Mobility of strontium-90 in alluvial groundwater, Los Alamos Canyon: Implications for pathway analysis" 8th TIE meeting, Santa Fe, 1995
14. K.J. Cantrell, P.F. Martin, and J.E. Szecody, "Clinoptilolite as an in-situ permeable barrier to strontium migration in ground water" pp 839-849, In-Situ Remediation: Scientific Basis for Current and Future Technologies, G.W. Gee and N.R. Wing (Eds.) Battelle Press, Columbus, Ohio, 1994
15. "Caesium cut from radioactive waste" C&EN p. 26 July 13, 1992
16. J.L. Conca, M.J. Apted, W. Zhou, R.C. Arthur and J.H. Kessler, " Flow barrier system for long-term high-level-waste isolation: experimental results." Nuclear technology, vol. 124, pp 88–100, October 1988
17. K. Watanabe, "Archeological evidence supports a preferred soil cap for LLRW disposal," Proc. of the Joint Int. Waste Conference, vol. 1, pp 567-571, 1989

# EVALUATION OF THE ANTHROPOGENIC RADIONUCLIDE CONCENTRATIONS IN SEDIMENTS AND FAUNA COLLECTED IN THE ARCTIC

D.W. EFURD, F.R. ROENSCH
*Nuclear and Radiochemistry,*
*Los Alamos National Laboratory,*
*Los Alamos, NM, 87545, U.S.A.*

W.C.T. INKRET
*Radiological Dose Assessment,*
*Los Alamos National Laboratory,*
*Los Alamos, NM, 87545, U.S.A.*

M.J. HAMEEDI
*National Oceanic and Atmospheric Administration,*
*Silver Spring, MD, 20910, U.S.A.*

C.F.V. MASON
*Nuclear Materials Technology,*
*Los Alamos National Laboratory,*
*Los Alamos, NM, 87545, U.S.A.*

## Abstract

This study was performed to establish a quality controlled data set about the levels of anthropogenic radionuclide activity in the environment and in selected biota in the Arctic. Between 1993 and 1997 sediment and marine and terrestrial biota samples were collected in Alaska and in the surrounding waters. These samples were analysed at Los Alamos National Laboratory to identify the presence of the anthropogenic radionuclides, $^{90}$Sr, $^{137}$Cs, $^{238}$Pu, $^{239+240}$Pu, and $^{241}$Am. One goal of this study was to determine the amounts and origins of anthropogenic radionuclides present in sediments. The anthropogenic radionuclide content of the sediments was predominantly the result of deposition of global fallout from nuclear explosions. No other sources of anthropogenic radionuclides could be conclusively identified. Biota samples of subsistence and ecological value were analysed for evidence of bioaccumulation of radionuclides. The measurements were combined with food consumption rates based on survey results for populations residing in three north Alaskan communities and on published, age-dependent ingestion coefficients. This action was carried out to estimate



*S.S. Hecker et al. (eds.), Nuclear Physical Methods in Radioecological Investigations of Nuclear Test Sites,* 99-107.
© 2000 *Kluwer Academic Publishers. Printed in the Netherlands.*

the potential radiological impacts from the consumption of traditional animal foods harvested in this region. No plutonium or americium was detected in any of the fauna samples. The results of this study indicate that the committed equivalent doses to adults from $^{90}Sr$ and $^{137}Cs$ owing to consumption of traditional food sources are consistent with currently accepted estimates of average doses received by adults in North America from atmospheric nuclear weapons testing fallout.

## 1. Methodology

To perform this study, researchers collected grab samples of sediments at various locations in the Arctic. Samples were collected as far to the east as Camden Bay and as far to the west as Amchitka Island. Figure 1 illustrates these sampling locations. The top 2-3 cm of each grab sample were reserved for radiochemical analyses. Whenever practical, marine samples were collected at the same locations. Land animals of importance to the diets of individuals living by subsistence were also collected.

*Figure 1.* A typical high resolution gamma ray spectrometer used to analyse the sediment and biota samples.

The samples were frozen and packed in dry ice for shipment to the Los Alamos National Laboratory (LANL). There, the samples were thawed, dried to constant weight, and ashed prior to analyses. The ashed samples were analysed by high-resolution gamma ray spectroscopy, the apparatus shown in Figure 1. After the high resolution gamma ray spectroscopy measurements were completed, the ashed samples were dissolved in strong mineral acids. Plutonium-238, $^{239}$Pu, $^{240}$Pu, $^{137}$Cs, and $^{90}$Sr were analysed from a single aliquot of each sample. Plutonium was separated and purified by ion exchange chromatography. The plutonium was electroplated onto platinum and the $^{238}$Pu and $^{239+240}$Pu concentrations were measured by alpha spectroscopy. The plutonium was then removed from the platinum disks and the $^{240}$Pu/$^{239}$Pu atom ratios were measured by thermal ionisation mass spectrometry (TIMS). Next, the iron was removed from the sediment samples by isopropyl ether extraction. The small amount of iron present in the biota was removed as the chlorocomplex by anion exchange. This was done to maximise the chemical yields for caesium and strontium. The caesium fractions were isolated by sorption on microcrystalline ammonium phosphomolybdate. Caesium was measured by low-level beta counting and high-resolution gamma ray spectroscopy. The strontium fractions were isolated as carbonates and purified by cation exchange chromatography. The $^{90}$Sr activity levels were determined by measuring the $^{90}$Y daughter using low-level beta counters. The $^{241}$Am content was measured on separate $\approx$ 5 g aliquots of ash or sediments. Americium was separated by co-precipitation with lanthanum fluoride and by a HDEHP (di-2-ethylhexylphosphoric acid in heptane) extraction. Final purification was accomplished by anion exchange chromatography. The americium was quantified by solid state alpha spectrometry. The detailed chemical separation and purification procedures have been published in Los Alamos National Laboratory report LA-13302-MS [1].

The high resolution gamma ray spectroscopy measurements were obtained by placing the ashed residues in counting vials 4 cm o.d. x 2.7 cm high and then counting for 3000 minutes, using an intrinsic germanium detector with an 81% efficiency relative to a 3 in x 3 in sodium iodide crystal. The resolution of the intrinsic germanium detector is 1.7 keV FWHM at 1.33 MeV. The efficiency of the detector is dependent on energy. We measured the efficiency versus energy, using multiline gamma standards that are traceable to the United States National Institute of Standards and Technology (NIST). We also fit the data to a semiempirical polynomial so that we could predict the efficiency at energies different from the calibration points. The separated caesium samples were counted for 1000 minutes in a well-type intrinsic germanium detector having a 26.5% efficiency relative to sodium iodide. The resolution of this detector is 2.4 keV FWHM at 1.33 MeV.

The TIMS used in this study to measure the $^{240}$Pu/$^{239}$Pu atom ratios is a modified version of the 30.5-cm radius, 90° deflection, thermal ionisation instrument developed at the former United States National Bureau of Standards (NBS) (now known as NIST). This spectrometer as shown in Figure 2 is equipped with a 17-stage electron multiplier with a gain of $10^8$, a discriminator-amplifier capable of supplying 1 V, 7-ns-wide output pulses when triggered by input pulses between 6 mV and 2 V, and a 250 MHz scaler. The dark current of the counting system is 0.05 counts per second. The system's overall linearity and deadtime were verified by analysing NBS' standard reference materials: U-100, U-500, and U-900. The data acquisition is automated with an IBM personal computer. The spectrometer is equipped with a sample turret that holds five surface ionisation diffusion controlled (SID) filaments.

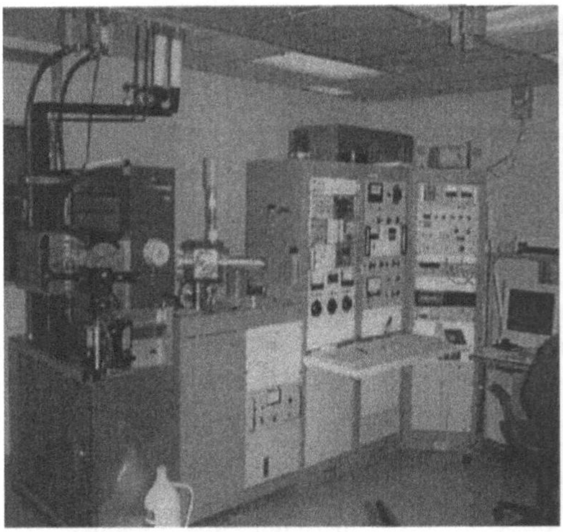

*Figure 2*. The thermal ionisation mass spectrometer (TIMS) used to measure the plutonium content and isotopic composition of the samples.

We also have an extensive quality assurance/quality control (QA/QC) programme to verify that the instrument measurement systems are functioning properly. Each gamma counting system has one or more standards that are counted each day. These data are then computer processed to check energy calibration, energy resolution, and overall detection efficiency. These data are then archived as well as plotted on running plots that are updated weekly for reference. Absolute detection efficiency checks are made whenever any out-of-normal incident occurs. Background counts are made whenever there are no samples being counted. The background data are also analysed to test for possible counter contamination.

The NIST Standard Reference Materials (SRM) for Environmental Radioactivity were processed concurrently as part of our quality assurance programme. These SRM served as the matrix-based, quality assurance standards for the sediment analyses. Table 1 lists the NIST-certified, matrix-based reference materials that we used during this investigation. Processing blanks were also analysed as part of our QA/QC programme. All NIST standards were sediments. Comparable ashed biota standards do not currently exist.

TABLE 1. NIST Matrix-Based QA/QC Samples

| | |
|---|---|
| SRM 4350B | Columbia River sediment |
| SRM 4353 | Rocky Flats soil Number 1 |
| SRM 4354 | Freshwater Lake sediment |
| SRM 4355 | Peruvian soil |

## 2. Measurement Results

Table 2 shows the representative plutonium contents and $^{240}Pu/^{239}Pu$ atom ratios measured in the sediments. The plutonium concentrations and $^{240}Pu/^{239}Pu$ atom ratios are consistent with those of global fallout. The average $^{240}Pu/^{239}Pu$ atom ratio measured in the sediments reported in Table 2 was 0.193 ± 0.011. Krey *et al.* [2] measured the $^{240}Pu/^{239}Pu$ atom ratio in global fallout to be 0.176 with a standard deviation of 0.014 and ranges from 0.12 to 0.21 around the world. The $^{238}Pu$ levels in the sediments were below our limits of detection.

TABLE 2. Pu-239/Pu-240 atom ratios and Pu-239 + Pu-240 concentrations in Arctic sediments

| Collection area | Date | Location | Pu-239/Pu-240 atom ratio | Pu $^{239+240}$ Bq/kg sediment |
|---|---|---|---|---|
| Beaufort Sea | 8/10/93 | 73:18N;167:06W | 0.187 ± 0.021 | 0.093 ± 0.028 |
| Beaufort Sea | 8/10/93 | 73:54N;168:30W | 0.204 ± 0.017 | 0.034 ± 0.007 |
| Beaufort Sea | 8/11/93 | 73:06N;166:60W | 0.186 ± 0.021 | 0.364 ± 0.032 |
| Beaufort Sea | 8/12/93 | 71:60N;165:21W | 0.173 ± 0.038 | 0.141 ± 0.016 |
| Beaufort Sea | 8/14/93 | 71:54N;154:46W | 0.179 ± 0.024 | 0.163 ± 0.05 |
| Bristol Bay | 7/18/94 | 54:67N;165:00W | 0.205 ± 0.006 | 0.352 ± 0.007 |
| Bristol Bay | 7/19/94 | 55:67N;163:67W | 0.202 ± 0.007 | 0.246 ± 0.006 |
| Bristol Bay | 7/21/94 | 56:17N;160:82W | 0.199 ± 0.011 | 0.208 ± 0.006 |
| Bristol Bay | 7/20/94 | 56:50N;162:01W | 0.216 ± 0.007 | 0.139 ± 0.003 |
| Bristol Bay | 7/21/94 | 57:00N;159:00W | 0.205 ± 0.013 | 0.121 ± 0.004 |
| Bristol Bay | 7/22/94 | 57:49N;160:01W | 0.193 ± 0.012 | 0.137 ± 0.005 |
| Bristol Bay | 7/21/94 | 58:00N;158:49W | 0.207 ± 0.012 | 0.069 ± 0.002 |
| Yukon-Kuskokwim | 7/24/94 | 60:67N;166:77W | 0.182 ± 0.011 | 0.114 ± 0.003 |
| Yukon-Kuskokwim | 7/25/94 | 61:70N;167.01W | 0.181 ± 0.007 | 0.092 ± 0.002 |
| Yukon-Kuskokwim | 7/25/94 | 62:67N;166:01W | 0.193 ± 0.009 | 0.071 ± 0.002 |
| Yukon-Kuskokwim | 7/26/94 | 63:63:N;163:00W | 0.199 ± 0.007 | 0.081 ± 0.002 |
| Yukon-Kuskokwim | 7/26/94 | 63:93N;162:01W | 0.194 ± 0.012 | 0.140 ± 0.004 |
| Norton Sound | 7/27/94 | 64:34N;165:36W | 0.188 ± 0.013 | 0.114 ± 0.004 |
| Norton Sound | 7/26/94 | 63:83N;165:17W | 0.178 ± 0.005 | 0.132 ± 0.002 |
| Norton Sound | 7/30/94 | 63:34N;166:50W | 0.195 ± 0.010 | 0.092 ± 0.003 |
| Norton Sound | 7/30/94 | 63:34N;167:84W | 0.181 ± 0.013 | 0.172 ± 0.006 |
| Norton Sound | 7/30/94 | 63:60N;169:02W | 0.201 ± 0.013 | 0.183 ± 0.006 |
| Norton Sound | 7/30/94 | 64:00N;168:00W | 0.193 ± 0.009 | 0.127 ± 0.003 |
| Norton Sound | 8/3/94 | 64:50N;166:67W | 0.199 ± 0.007 | 0.086 ± 0.002 |
| Norton Sound | 8/2/94 | 65:26N;167:68W | 0.193 ± 0.011 | 0.076 ± 0.002 |

A subset of the samples reported in Table 2 was analysed for $^{90}Sr$, $^{137}Cs$, $^{238}Pu$, and $^{241}Am$. The average $^{241}Am/^{239+240}Pu$ alpha activity ratio in these sediments was 0.48 ± 0.23. This value is not statistically different from the $^{241}Am/^{239+240}Pu$ alpha activity ratio of 0.33 that is present in global fallout. The average $^{239+240}Pu/^{137}Cs$ activity ratio determined in the sediment samples was 0.05 ± 0.02.

The $^{90}Sr$ concentrations in the sediments were significantly lower than the $^{137}Cs$ concentrations. The amount of $^{90}Sr$ relative to $^{137}Cs$ and $^{239+240}Pu$ in the samples analysed is depleted relative to global fallout. We can only speculate that the reason for this anomaly is the fact that strontium is water soluble. Caesium, plutonium, and americium are much more insoluble than strontium.

The anthropogenic radionuclide content of the sediments was predominantly the result of deposition of global fallout. No other sources of anthropogenic radionuclides could be conclusively identified. The preliminary analyses of samples collected in 1997 show comparable results.

Biota samples of subsistence and ecological value were analysed for evidence of bioaccumulation of radionuclides. Samples were analysed for $^{90}Sr$, $^{137}Cs$, $^{238}Pu$, $^{239+240}Pu$, and $^{241}Am$. Though $^{90}Sr$ and $^{137}Cs$ were detected in many of the samples, no plutonium or americium was detected in any of the fauna samples.

Table 3 illustrates the typical results for fauna collected from northern Alaska and the Beaufort Sea. In this study the $^{90}Sr$ activity concentrations in fish collected from the Beaufort Sea were of the order of from 0.1 to 0.5 Bq/kg. This level of concentration produces an estimated 0.13 μSv (committed effective dose) per kilogram of fish consumed from the Beaufort Sea.

TABLE 3. Measured radionuclide concentrations (Bq/kg, wet weight) in food sources collected from Northern Alaska and the Beaufort Sea

| Resource | Sr-90 | Cs-137 |
|---|---|---|
| Broad whitefish | 0.11 ± 0.06 | 0.06 ± 0.01 |
| Arctic char | 0.12 ± 0.08 | < 0.04 |
| Arctic cod | < 0.05 | 1.35 ± 0.05 |
| Arctic cisco | 0.47 ± 0.24 | 0.62 ± 0.12 |
| Least cisco | Not measured | 0.26 ± 0.09 |
| Dolly varden | 0.13 ± 0.03 | 0.17 ± 0.07 |
| Chum salmon | < 0.04 | 0.11 ± 0.03 |
| Bowhead whale | 0.015 ± 0.003 | 0.011 ± 0.03 |
| Bearded seal | < 0.02 | 0.17 ± 0.14 |
| Polar bear | 0.03 ± 0.03 | 0.17 ± 0.14 |
| King eider | < 0.001 | < 0.13 |
| Caribou | Not measured | 9.2 ± 1.1 |

Note that the caribou $^{137}Cs$ concentrations are much larger relative to the other animal tissues reported in Table 3. In arctic and subarctic environments, the lichen-to-caribou-to-man food chain constitutes a well documented pathway for concentration of a radioactive species in a human food source.

The $^{137}$Cs activity concentrations in fish collected in the Beaufort Sea and used in this study were of the order of from 0.1 to 1 Bq/kg. This level of concentration is approximately 50% of the values measured from fish taken from the Barents Sea (0.2-3 Bq/kg) and significantly lower than samples collected from the Kara Sea (11 to 26Bq/kg).

Table 4 summarises the average intake values for the food sources listed in Table 3, as derived from the Alaska Department of Fish and Game (ADFG) survey results. The uncertainties listed in Table 4 are calculated as standard deviation of the average annual per capita results; these uncertainties are not based on individual variations in intake. The results reported without uncertainties are based on only one or two annual surveys.

TABLE 4. Average per capita, annual consumption rates of traditional marine and terrestrial foodstuffs listed in Table 3 for Barrow, Kaktovik, and Nuiqsut, Alaska

| Resource | Barrow (kg/capita/year) | Kaktovik (kg/capita/year) | Nuiqsut (kg/capita/year) |
|---|---|---|---|
| Broad whitefish | 5.96 ± 5.24 | 0 | 41.42 |
| Arctic char | 0.26 ± 0.26 | 22.99 ± 12.69 | 0 |
| Actic cod | 0.51 ± 0.17 | 0.15 | 0.01 |
| Arctic cisco | 0.21 ± 0.16 | 3.31 ± 3.03 | 36.5 |
| Least cisco | 0.66 ± 0.34 | 0.82 | 4.1 |
| Dolley varden | 0 | 0 | 0 |
| Chum salmon | 0.17 ± 0.17 | 0 | 1.05 |
| Bowhead whale | 32.4 ± 22.4 | 137.5 | 52.52 |
| Bearded seal | 3.98 ± 1.96 | 8.65 ± 1.61 | 2.16 |
| Polar bear | 1.39 ± 1.35 | 2.47 ± 0.85 | 0 |
| King eider | 0.02 ± 0.02 | 0.79 ± 0.39 | 0.99 |
| Caribou | 23.4 ± 9.2 | 47.01 ± 13.42 | 72.24 |

The results of the radioactivity concentration measurements listed in Table 3 were combined with the food consumption rate, collected and compiled by ADFG between 1980 and 1997, to estimate ingestion intake rates of the detected radioactive species in the three Alaskan communities of Barrow, Kaktovik, and Nuiquist. The ADFG compilations focus on consumption of marine and terrestrial food sources harvested for subsistence in Alaskan communities.

Age-specific annual radioactivity intakes from traditional food sources were estimated by combining the ADFG data, the measured radionuclide concentrations in Table 3, and the age-dependent fractions of daily intake.

The committed effective dose from ingestion of $^{137}$Cs measured in marine and terrestrial food sources listed in Table 5 range from 2.2 ± 0.3 to 9.1 ± 1.1 μSv, for an adult.

TABLE 5. Age-dependent committed equivalent dose estimates from $^{90}Sr$ and $^{137}Cs$ after a single year of consumption of marine and terrestrial food sources as listed in Tables 1 and 3 for three Alaskan communities

| Median age (years) | Bone surface equivalent dose (μSv) | Effective dose (μSv) | Effective dose (μSv) |
|---|---|---|---|
| Barrow | $^{90}Sr$ | | $^{137}Cs$ |
| 0.25 | 0.11 ± 0.03 | 0.010 ± 0.003 | 0.27 ± 0.04 |
| 1 | 0.10 ± 0.03 | 0.010 ± 0.003 | 0.46 ± 0.07 |
| 5 | 0.15 ± 0.04 | 0.011 ± 0.003 | 0.63 ± 0.10 |
| 10 | 0.26 ± 0.07 | 0.014 ± 0.004 | 0.69 ± 0.10 |
| 15 | 0.74 ± 0.19 | 0.033 ± 0.009 | 1.7 ± 0.2 |
| >18 (adult) | 0.23 ± 0.06 | 0.016 ± 0.004 | 2.2 ± 0.3 |
| Kaktovik | | | |
| 0.25 | 2.43 ± 0.48 | 0.24 ± 0.05 | 0.73 ± 0.21 |
| 1 | 2.43 ± 0.48 | 0.24 ± 0.05 | 1.25 ± 0.37 |
| 5 | 3.58 ± 0.71 | 0.26 ± 0.05 | 1.72 ± 0.51 |
| 10 | 6.2 ± 1.2 | 0.34 ± 0.07 | 1.87 ± 0.55 |
| 15 | 17 ± 3 | 0.76 ± 0.15 | 4.59 ± 1.35 |
| >18 (adult) | 5.5 ± 1.1 | 0.37 ± 0.07 | 5.91 ± 1.74 |
| Nuiqsut | | | |
| 0.25 | 4.74 ± 1.66 | 0.47 ± 0.17 | 1.11 ± 0.13 |
| 1 | 4.71 ± 1.65 | 0.46 ± 0.16 | 1.92 ± 0.22 |
| 5 | 6.93 ± 2.43 | 0.51 ± 1.1 | 0.37 ± 0.07 |
| 10 | 11.9 ± 4.17 | 0.65 ± 0.23 | 2.87 ± 0.34 |
| 15 | 33.4 ± 11.7 | 1.47 ± 0.51 | 7.02 ± 0.82 |
| >18 (adult) | 10.6 ± 3.7 | 0.72 ± 0.25 | 9.06 ± 1.06 |

The National Council on Radiation Protection (NCRP) has published a value of 150μSv from chronic intakes of environmental $^{137}Cs$ over a 30 year period (from 1970 to 2000) for an adult in the United States. Assuming an exponential decrease with a half-time of 30 years and no other competing pathways to remove caesium from the food chain, the committed dose for a single year of uniform daily intake in 1993 would be on the order of 4μSv.

In this study, the $^{90}Sr$ activity concentrations in fish in the Beaufort Sea were of the order of from 0.1 to 0.5 Bq/kg. These concentrations result in 0.13 μSv (committed bone surface equivalent dose) and 0.009 μSv (committed effective dose) per kg of fish consumed from the Beaufort Sea.

The NCRP has estimated a committed bone surface dose of 650 μSv from chronic intake of environmental $^{90}Sr$ over a 30-year period, from 1970-2000 for adults in the United States of America. Assuming an exponential decrease with a half-time of 30 years and no other competing pathways to remove strontium from the food chain, the committed dose for a single year of uniform daily intake in 1993 would be of the order of 14 μSv. The committed bone surface equivalent dose to adults from a single year of intake of $^{90}Sr$, estimated in this study, is 0.23 ± 0.06 μSv to 10.6 ± 3.7 μSv.

The results presented here indicate that the internal radiation dose from intake of traditional marine and terrestrial animal food sources in northern Alaska are consistent with published estimates of doses to adults in the U. S. population due to natural background and fallout from atmospheric nuclear weapons testing. It is emphasised that the doses reported here are based on a subset of subsistence intakes; doses from total annual ingestion could be 10% to 30% higher.

## 3. Conclusions

The anthropogenic radionuclide content of the sediments was predominantly the result of deposition of global fallout. No other sources of anthropogenic radionuclides could be conclusively identified.

The results of this study indicate that the committed equivalent doses to adults from $^{90}Sr$ and $^{137}Cs$ owing to consumption of traditional food sources are consistent with currently accepted estimates of average doses to adults in North America, resulting from atmospheric nuclear weapons testing fallout.

## 4. Acknowledgements

Funding for this study was provided by the National Oceanic and Atmospheric Administration and the U.S. Department of Energy. The National Biological Service and the North Slope Borough's Department of Wildlife Management provided several sediment and biological samples. The authors wish to thank Don Rokop, Harold Poths, Phil Hemberger, and John Chamberlin of Los Alamos National Laboratory for providing mass spectometric analyses support.

## References

1. Efurd, D. W., et al. "Evaluation of the Anthropogenic Radionuclide Concentrations in Sediments and Fauna Collected in the Beaufort Sea and Northern Alaska," Los Alamos National Laboratory, Report LA-13302-MS (February 1997).
2. P.W. Krey, E.P. Hardy, C. Pachucki, F. Rourke, J.Coluzza aaand W.K. Benson, "Mass Isotopic Composition of Global Fall-Out Plutonium in Soil," in *Proceedings of a Symposium on "Transuranium Nuclides in the Environment,"* San Francisco, 17-21 November 1975, International Atomic Energy Agency report STI/PUB/410 (1976).

# RADIOLOGICAL ENVIRONMENTAL RESEARCH CONDUCTED AT THE SITE OF THE SINKING OF THE "KOMSOMOLETS" NUCLEAR-POWERED SUBMARINE

I.B. ASTAKHOV, A.A. DRUZHININ, N.G. KRYLOV,
A.M. KOROCHKIN, M. YU. MAKSIMOV
*Russian Federal Nuclear Centre,*
*All-Russian Scientific Institute of Experimental Physics,*
*Sarov, Russia*

## 1. Introduction

On April 7, 1989, in the Norwegian Sea, the nuclear-powered submarine (NS) "Komsomolets" sank [1]. This was not the first submarine to sink with atomic facilities and nuclear weapons on board. For example, on May 22, 1968, in the Atlantic Ocean several hundreds of miles from the Azores, the American submarine "Scorpion" sank with two Mark 45 Astor torpedoes with nuclear warheads on board [2]. Other nuclear-powered submarine wrecks have also occurred. However, the Komsomolets sinking caused the greatest world-wide repercussions.

Almost immediately after the wreck of the Komsomolets, the then Soviet Government ordered the leading scientific institutes and organisations including the Russian Federal Nuclear Centre of the All-Russian Scientific Institute of Experimental Physics (RFNC-VNIIEF) to develop and then perform the activities necessary for eliminating the possible radiation hazards from the atomic reactor and torpedoes with nuclear warheads on the wrecked submarine located at the depth of ~1700 m.

Scientific marine expeditions carried out between 1989 and 1995 carefully investigated the Komsomolets submarine with deep-sea inhabited apparatus (DIA) "MIR-1" and "MIR-2" of the Oceanology Institute of the Russian Academy of Science. Afterwards, these hazards were reliably isolated from the environment through the use of special means and materials [3].

The present report describes some radiation monitoring investigations performed in this part of the Norwegian Sea. These inquiries were carried out by RFNC-VNIIEF specialists in on-site scientific expeditions as well as by others in the scientific laboratories of the Institute.

Individual research results were reported at previous Russian-American meetings held in 1997 [4] and in 1998 [5].



*S.S. Hecker et al. (eds.), Nuclear Physical Methods in Radioecological Investigations of Nuclear Test Sites,* 109-114.
© 2000 *Kluwer Academic Publishers. Printed in the Netherlands.*

## 2. Experimental Research

Radiation monitoring of the Komsomolets submarine was conditioned by the radiation situation control requirements in effect when one is conducting the work on the submarine study and on its confinement from the environment. Monitoring was also conditioned by the need to obtain data on the state of reactor and nuclear torpedoes, information required to forecast the present situation in the Norwegian Sea region.

The major attention of our Institute concerning the radiation monitoring of the area where the Komsomolets sank focused on obtaining experimental information on the isotopic composition and concentrations of fissile materials (uranium and plutonium) in the environment, the known amounts of which entered into the composition of the nuclear charges and atomic reactor located on board the submarine. The difficulty of this task resulted from the human activity that caused the plutonium to be scattered over the whole Earth and, correspondingly, around the region of the sinking [6].

Simultaneously, traditional radiation monitoring was conducted. Namely, the gamma-ray dose rate near the wrecked NS was registered and the spectral composition of this gamma-radiation was studied. Because background plutonium concentrations in sea water are characterised by extremely low values, the most sensitive methods of radiochemical analysis were applied in this research.

### 2.1. AN OBJECT UNDER STUDY AND EXPERIMENTAL TECHNIQUES

Multiple samples of water, suspensions, and bottom sediments were taken with the aid of special sampling devices submerged from onboard the scientific research ship “Akademik Mstislav Keldysh” and installed on the submerged “MIR-1” and “MIR-2” DIAs.

To conduct analysis onboard ship, researchers used mobile radiometric and spectrometric equipment as well as proximate radiochemical techniques of releasing and refining uranium and plutonium. During marine expeditions the following instruments were used:

- a mobile MI 1201B mass-spectrometer variant,
- a GAR-01 deep sea semiconductor gamma-spectrometer, and
- a low-background, semiconductor alpha-spectrometer “Aladin,” as well as some other devices.

The radiochemical technique that was used for uranium detection [7] consisted of the chemical release from samples under analysis and the subsequent performance of mass-spectral measurements of isotopic composition with the use of the known amount of a high-enriched uranium-233 preparation as an isotopic mark (99.9728% of mass.).

This radiochemical technique of plutonium [8] detection consisted of its chemical release from the samples under analysis and subsequent alpha-spectral (when determining concentration) or mass-spectral measurements (when determining isotopic composition) with the use of isotopically pure plutonium-242. The known amount of this pure plutonium-242 was added to the sample under analysis before the beginning of chemical operations. Impurities of other plutonium isotopes in the mark included: $^{239}Pu<8\cdot10^{-6}$; $^{240}Pu$-$1.6\cdot10^{-4}$; $^{238}Pu$-$3\cdot10^{-4}$ with regard to the main $^{242}Pu$ isotope activity.

## 2.2. MEASUREMENTS RESULTS AND DISCUSSION

Table 1 gives the results of determining concentrations and isotopic composition of uranium and plutonium in the samples of water and bottom sediments, taken with the help of the "MIR" DIA near the Komsomolets, as well as in the torpedo compartment inside the destroyed submarine bow.

TABLE 1. Results of measurements of concentrations and isotopic composition of uranium and plutonium in samples of water and bottom ground taken inside and near the NS

| Year | Uranium | | | | Plutonium | | | | |
|---|---|---|---|---|---|---|---|---|---|
| | Concentration | | $^{235}U/^{238}U$ ($^*10^{-3}$) | | Concentration | | | $^{238}Pu$ / $^{239,240}Pu$ Bq/Bq | $^{240}Pu$ / $^{239}Pu$ at/at |
| | Water mkg/l | Ground mkg/g | Water at/at | Ground at/at | Water mBq/m³ | Suspension mBq/m³ | Ground mBq/kg | | |
| 1994 | 3.14 | 2.56 | 7.22 | 7.24 | ~ 82 | 13±3 | 220±20 | 0.2±0.1 | 0.19±0.03 |
| | 3.25 | 2.25 | 7.24 | 7.26 | ~ 140 | 7±2 | | 0.3±0.2 | 0.16±0.04 |
| | 3.23 | 4.00 | 7.26 | 7.25 | - | 32±3 | | 0.3±0.2 | 0.19±0.03 |
| | 3.08 | 1.75 | 7.20 | 7.24 | ~ 150 | | | 0.22±0.0 | |
| | 3.07 | 1.73 | 7.23 | | ~ 150 | | | 4 | |
| | 3.03 | 2.17 | 7.24 | | | | | | |
| | 3.31 | 2.30 | 7.25 | | | | | | |
| | 3.13 | | 7.22 | | | | | | |
| | 3.20 | | 7.23 | | | | | | |
| | | | 7.26 | | | | | | |
| 1995 | 4.62 | | 7.25 | | 40±10 | 7±3 | | 0.2±0.1 | 0.18±0.04 |
| | 3.43 | | 7.24 | | 20±10 | 4.2±0.8 | | 0.1±0.05 | 0.18±0.05 |
| | 3.45 | - | 7.24 | - | 31±5 | 5.3±1.0 | - | 0.2±0.1 | |
| | 3.41 | | 7.23 | | | | | 0.2±0.1 | |
| | 2.32 | | 7.22 | | | | | | |
| | 3.27 | | 7.26 | | | | | | |
| | | | 7.15 | | | | | | |
| | | | 7.24 | | | | | | |

Notes:
1) The term "water" indicates a summary sample of water and suspension.
2) Each separate number corresponds to one sample.
3) The uncertainty of measurements of the ratio $^{235}U/^{238}U$ constitutes $\pm1\%(2\sigma)$; uranium concentrations constitute $\pm2\%(2\sigma)$.

The excess concentrations in water samples taken from the NS in 1994, as compared to the subsequent testing in 1995, is conditioned by uncertainties of the 1994 sampling. In those 1994 water samples, there was a rather high amount of sea bottom sediments containing plutonium concentration significantly higher than those in water.

Let us consider the data obtained from the point of view of their correspondence to the known values of the natural background of these nuclides in the sea and the sea bottoms.

### 2.2.1. *Uranium*

It is well known that uranium in a soluble form is a constituent of marine water. Its concentration, equal to 3.3±0.4 mkg/l, is practically constant for all depths and regions of the world's oceans [9]. Isotopic ratios in the uranium of natural enrichment—$^{235}U/^{238}U$ and $^{234}U/^{238}U$, equalling, relatively, $(7.25\pm0.05)\cdot10^{-3}$ at/at and $(5.5\pm0.2)\cdot10^{-5}$ at/at, are also practically constant over the whole ocean, independent of the place of sampling. Comparing these worldwide values with those measured in the samples, we can conclude that they fall within the framework of experimental uncertainties. Therefore, structural uranium, which was onboard the Komsomolets, was not found in the environment.

The lowest threshold of structural uranium detection in water samples can be described in the following way. If a nonsignificant yield of weapons uranium from nuclear charges into the internal submarine space (1-st, 2-nd compartments, space~$10^3$ $m^3$) did occur, it would be registered by a relative increase of the ratio $^{235}U/^{238}U$ of 3%. That is, instead of $7.25\cdot10^{-3}$, it would be $7.50\cdot10^{-3}$. Such an amount is equivalent to the accumulation of ~50÷100 mg weapons uranium-235 in this submarine space.

Estimates show that even in case of a total release of the weapons uranium into the environment, no significant increase of uranium-235 concentration in water would occur due to a considerable content of natural uranium in the sea and a corresponding dilution because of the process of bottom-water transfer (speed of the order of 10 cm/s [3]).

### 2.2.2. *Plutonium*

Plutonium, which does not occur naturally in nature, was injected into the Earth, mostly as a result of atmospheric nuclear weapons tests and space vehicle accidents [10]. It is known that plutonium concentration in the oceans worldwide is not of a constant magnitude, as compared to uranium, but varies within the wide limits (~1÷100 $mBq/m^3$) depending on the ocean's region, its depth, and its distance from atomic industry plants, weapons proving grounds, etc. [11]. We possess no specific data on global plutonium content in the Norwegian Sea. We can compare the obtained plutonium concentration in water samples taken from the bow of the wrecked submarine (where nuclear torpedoes were located) with the background plutonium concentration values measured at a distance of several tens of kilometres from the submarine wreck region. The results of such a comparison are given in Table 2.

Table 2. Plutonium concentration in water samples from the Komsomolets bow and in samples taken from the region around the wreck

| Sampling place | Sampling level ( m ) | Pu-239, Pu-240 concentration ($mBq/m^3$) | | Year |
|---|---|---|---|---|
| | | In sample | Average value | |
| First compartment of the Komsomolets | 1690 | 40 ± 10 | | |
| | 1690 | 20 ± 10 | 31 ± 8 | 1995 |
| | 1690 | 31 ± 5 | | |
| Background stations of the near and far zones | 1200 | 13 ± 4 | | |
| | 1700 | 8.6 ± 3 | 10 ± 3 | 1991 |
| | 1700 | 12 ± 4 | | |
| | 1700 | 9 ± 3 | | |
| | 1632 | 47 ± 15 | | |
| | 1640 | 9 ± 3 | 18 ± 5 | 1992 |
| | 1800 | 11 ± 4 | | |
| | 1655 | 44 ± 14 | | |
| | 1630 | 40 ± 13 | 22 ± 8 | 1993 |
| | 1655 | 16 ± 5 | | |
| | 2440 | 23 ± 7 | | |
| | 1730 | 20 ± 6 | 18 ± 5 | 1994 |
| | 1980 | 15 ± 5 | | |

It is clear that the plutonium concentration in water samples taken inside and near the wrecked NS is practically equal to that taken from the wreck region. From the above data in Table 2, we can conclude that there is no plutonium release from the wrecked nuclear-powered Komsomolets submarine into the environment.

The same conclusion follows from the data analysis of isotopic plutonium. As we can see from Table 1, the ratio $^{240}Pu/^{239}Pu$ changes in different samples from 0.16 to 0.19. The average value constitutes 0.183±0.016 at/at, and within the limits of measurement errors coincide with the known value for background plutonium, equal to 0.18±0.02 at/at [12]. This value practically remains the same for every ocean region, excluding regions that are close to nuclear weapons proving grounds [13,14]. Plutonium from the nuclear reactor and the torpedoes on board the Komsomolets is characterised by a $^{240}Pu/^{239}Pu$ ratio magnitude that is several times lower than 0.18. That is why the summary ratio (global plutonium plus nuclear-powered submarine plutonium) would be characterised by an appreciably lower value than that measured in samples—0.18±0.03 at/at. Nuclear-powered submarine plutonium action on another isotopic plutonium ratio $^{238}Pu/^{239,240}Pu$ is insignificant. And though this ratio is measured with an appreciably lower accuracy than the ratio $^{240}Pu/^{239}Pu$, its magnitude exceeds essentially the value that is characteristic for nuclear-powered submarine plutonium.

Table 3 provides the given values of isotopic relations and concentrations of uranium and plutonium in the sea water. For comparison, the values of the world's oceans background are also presented. These values are characteristic of the background uranium and plutonium contained in this Norwegian Sea region before the Komsomolets submarine wreck.

Table 3. Comparison of the experimental and background values

| Parameter | Background values | Measured in water samples from the NS |
|---|---|---|
| Uranium concentration, mkg/l | $3.3 \pm 0.4$ | $3.31 \pm 0.08$ |
| $^{234}U/^{238}U$, at/at | $(7.25 \pm 0.05) \cdot 10^{-3}$ | $(7.25 \pm 0.03) \cdot 10^{-3}$ |
| $^{234}U/^{238}U$, at/at | $(5.5 \pm 0.2) \cdot 10^{-5}$ | $(4.4 \div 5.9) \cdot 10^{-5}$ |
| Plutonium concentration, $mBq/m^3$ | $1 + 100$ | $20 + 150$ |
| $^{240}Pu/^{239}Pu$, at/at | $0.18 \pm 0.02$ | $0.183 \pm 0.016$ |
| $^{238}Pu/^{239,240}Pu$, Bq/Bq | $0.1 + 0.2$ | $0.2 \pm 0.1$ |
| Suspensions/Water, % | $10 + 30$ | $15 \pm 3$ |

Comparing these data with each other, we see no significant differences. On this basis, we can ensure the absence of fissile materials yield (uranium, plutonium) from the wreck into the Norwegian Sea.

## References


1. The truth on the ocean bottom. Corr. S.Zigunenko, magaz."Tekhnika molod'ozhi," N10, 1990.
2. Sheldon R.B., Michne J.D., (1993) Deep sea radiological environmental monitoring conducted at the site of the nuclear-powered submarine "Scorpion" sinking, KAPL-4749, KAPL, Inc. Knolls Atomic Power Laboratory, New York.
3. Oceanology research and subwater engineering work at the site of the nuclear-powered submarine "Komsomolets" sinking, edited by Vinogradov M.Ye., Sagalevich A.M., Khetagurov S.V., M., Nauka, 1996.
4. Astakhov I.B., Vesnovski S.P., Gorbachev V.M., Druzhinin A.A., Korochkin A.M., Kravchenko S.G., Krylov N.G., Maksimov M.Yu., Mikhailov B.P., Shapovalov V.I., (1997) Hardware methodical complex for radiological environmental monitoring, report for the Russian-American meeting, Livermore, California, February 18-21.
5. Barkanov B.P., Gorbachev V.M., Larionov S.A., Man'ko A.I., Shapovalov V.I., (1998) Technologies of eliminating the consequences of the wreck with submerged nuclear weapons on the basis of experience of the "Komsomolets" nuclear-powered submarine, report for the Russian-American workshop on eliminating the consequences of nuclear weapons wrecks, Russia, Moscow, May 4-8.
6. Cochran J.K., Livingston N.D., Hirschlerh D.J., Surprenant L.D., (1987) Natural and thropogenic radionuclide distributions in the northwest Atlantic Ocean, Earth and Planetary Science Letters, 84, p.135.
7. Grindler J.E., (1962) The radiochemistry of Uranium, NAS NS 3050.
8. Buesseler K.O., (1987) The mass-spectrometric Determination of Fallout $^{239}Pu$ and $^{240}Pu$ in Marine Samples, J.Environ. Radioactivity 5, p.425.
9. Dreik Ch., Imbri G., Turekian K., (1982) Ocean by itself and for us, M., Progress.
10. Hardy E.P., Krey P.W., Volchok H.L., (1973) Global Inventory and Distribution of Fallout Plutonium, Nature, v.241, February 16, p.444.
11. Bowen V.T., Noshkin V.E., Livingston H.D., Volchok H.L., (1980) Earth and Planetary Science Letters, 49, p.411.
12. M-Sakanoue, Radiochimical Acta, v.42, 1987, p.103-112.
13. Krey P.W. et al., (1976) Transuranie Nuclides in the Environment, IAEA, p.671.
14. Bourlat Y., Milles-Lacroix J.C., Nazard R., (1994) Low-Level measurements of Actinides and Long-Lived Radionuclides in Biological and Environmental Samples, 5-th conference, Japan, July 10-15.

# METHODS OF PLUTONIUM, AMERICIUM, AND STRONTIUM DETERMINATION WITH PRELIMINARY RADIOCHEMICAL SEPARATION

# DETERMINATION OF RADIONUCLIDE SPECIES IN NATURAL WATERS, SOILS, AND SEDIMENTS

D. OUGHTON, B. SALBU, L. SKIPPERUD
*Department of Chemistry and Biotechnology,*
*P.O. Box 5026, Agricultural University of Norway, 1432 Aas Norway*
*email: deborah.oughton@ikb.nlh.no*

## Abstract

The transfer and mobility of radionuclides in the environment are influenced by their physico-chemical forms. Thus scientists need analytical techniques that can separate different species of radionuclides, as well as low-level techniques for measuring radionuclides in different fractions. This paper, which presents an overview of the methods used at the authors' laboratory, first focuses on speciation techniques such as ultrafiltration, chemical extraction, electron microscopy, and synchrotron radiation-based, solid-state techniques, and second, on low -level measurement of Pu-isotopes using accelerator mass spectrometry (AMS). Though AMS is an established technique for the ultra-low level determination of long-lived radionuclides, it has only recently been applied for the determination of Pu-isotopes. Case studies to illustrate the application of these techniques include the study of contamination in the exclusion zone around the Chernobyl Nuclear Power Plant and at the Mayak plutonium production plant in the Former Soviet Union.

## 1. Introduction

Scientists studying the migration of radionuclides need information on the physico-chemical forms of contaminants. The speciation of radionuclides in water, soil and sediments determines their transfer and mobility within ecosystems [1]. Biological availability of radionuclides is also dependent upon the physico-chemical form, while changes in speciation due to transformation processes and weathering add a kinetic, time-dependent dimension to distribution models (Fig. 1). Thus, analytical challenges to scientists include the development and application of techniques capable of separating these species and also methods of low-level radionuclide measurement. Mass spectrometry measurement techniques have the added advantage of giving information on isotope ratios, which can be used to identify sources and trace the migration of contaminated waters and sediments. This paper will review some of the methods used at



*S.S. Hecker et al. (eds.), Nuclear Physical Methods in Radioecological Investigations of Nuclear Test Sites*, 117-125.
© 2000 *Kluwer Academic Publishers. Printed in the Netherlands.*

the authors' laboratory, focusing first on speciation techniques such as ultrafiltration, chemical extraction, electron microscopy, and on synchrotron radiation-based studies, and second on accelerator mass spectrometry (AMS) for determining plutonium isotopes.

To illustrate the application of these techniques, short case studies will include investigations of contamination in the exclusion zone around the Chernobyl Nuclear Power Plant and at the Mayak plutonium production plant in the Former Soviet Union. At Chernobyl, the association of radionuclides with uranium oxide particles and changes in speciation and mobility with particle weathering can be contrasted with the behaviour of radionuclides at Mayak, where the majority of the radionuclides was released as relatively mobile species. Since the situation at Mayak is further complicated by different sources, data on Pu isotope ratios can provide useful information on source term. The analytical methods should be of relevance for investigations of bomb test sites.

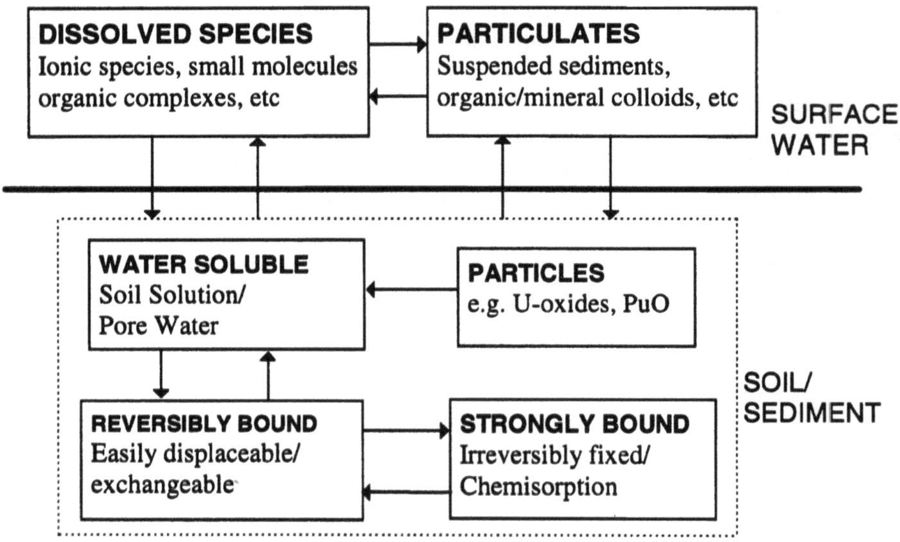


*Figure 1.* Examples of different physico-chemical forms of radionuclides in sediment water and soil-water systems.

## 2. Speciation Techniques

### 2.1. SIZE AND CHARGE FRACTIONATION

Size fractionation of species in waters can be carried out at a variety of molecular weight and size cut-offs. Filtration at 0.45 or 0.22 μm is standard for most water sampling. Charge fractionation can be carried out using a specific anion or cation exchange columns. However, standard filtration does not give information on the colloidal fraction of radionuclides in waters. This level of sampling requires ultrafiltration, usually at well-defined molecular cut-offs of between 1 to 30 kDa. A number of different techniques are available, but hollow-fibre filtration, using cross-flow membranes, has been shown to give reliable and reproducible results [2,3]. Whenever possible, fractionation should be carried out *in situ*, or as soon as possible after collection, since storage of water samples can lead to both a change in distribution of species and possibly a loss of radionuclides from solution owing to sorption and precipitation.

## 2.2. VALENCY/OXIDATION STATE

Actinides such as Pu, Np, and U can exist in a variety of oxidation states in natural waters, which in turn influence migration and uptake into organisms. Pu (V) and Pu (VI) or organically-bound Pu species tend to be more mobile, having soil-water and sediment-water distribution coefficients ( $K_d$, ml/g) of an order of magnitude lower than Pu (III) and Pu (IV) [4,5]. Thus, both the rate at which dissolved Pu species are removed from water to soil or sediment, and the eventual partitioning between the solid and dissolved species, vary as a function of the Pu oxidation state (Fig. 2). The established technique for separating plutonium species in waters is to use selective precipitation techniques, usually involving co-precipitation with $NdF_3$ or $LaF_2$ which carries $Pu^{3+}/Pu^{4+}$, leaving the higher oxidation state fraction $PuO_2^{+}/PuO_2^{2+}$ in solution [4].

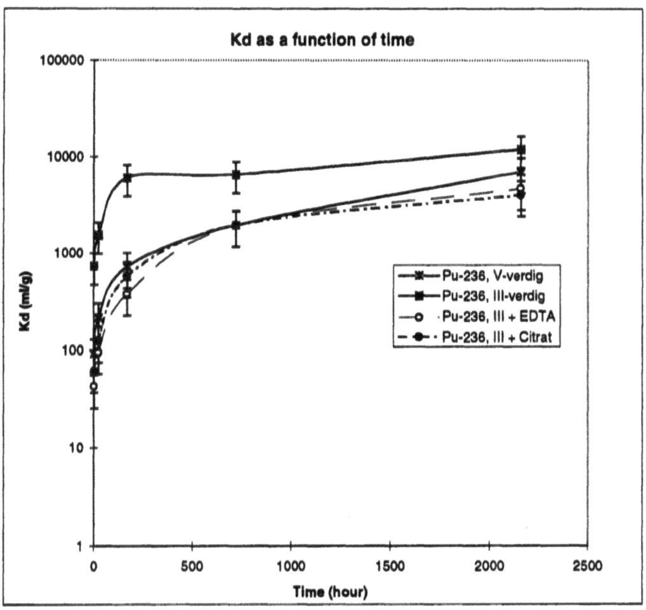


*Figure 2.* Influence of time and the Pu-oxidation state on the $K_d$ in a soil-water system [6].

## 2.3. SOIL AND SEDIMENT EXTRACTION

Soils and sediments often represent the largest inventories of radionuclides in the environment, and remobilisation can be a major source of radionuclides to surface and groundwaters [1,6]. The degree of that remobilization will depend on the physico-chemical form of the radionuclide as well as environmental variables such as pH, ionic strength and redox. Chemical extraction techniques are the most common method of studying radionuclide speciation in soils and sediments [7-9]. Although operationally defined, the methods are particularly valuable when combined with investigations of the change in speciation over time, including tracer studies and comparisons between radionuclides and stable analogues.

The comparison of the mobility of radionuclides in soils from Chernobyl and Mayak is a good example of the application of sequential extraction techniques. Mayak was established in the late 1940s to produce plutonium for the Soviet nuclear weapons programme. The site, located at the head of the Techa River in the Southern Urals, comprises military and civil reactors, as well as reprocessing and metallurgical plants [10]. Accidental and routine releases of radioactive waste, most notably the direct discharge of intermediate level waste to the Techa River (1949-1956) and Lake Karachay (1951-), as well as the Kyshtym accident, an explosion in a high-level waste storage tank in 1957, have caused severe contamination of the surrounding areas. Sequential extraction of soils and sediments from the Mayak area showed that Sr-90 was present in readily exchangeable forms, whereas Cs-137 and Pu-isotopes were strongly-bound to soil components [Fig. 3(a)]. Today, the major source of radionuclides to the surrounding area is believed to be remobilization from previously contaminated soils and sediments. Seepage through the beds of industrial waste storage reservoirs is contaminating groundwater; contaminated sediments are transported downstream by the Techa River waters during flooding; and chemically mobile radionuclides like $^{90}$Sr are leached out of contaminated soils.

At Chernobyl, much of the deposited Sr-90 and actinides in the near zone areas were associated with uranium oxide particles [12,13]. Sequential extraction of soils collected at various times after the 1986 accident showed that most of the radionuclides, including Sr-90, were initially present in a non-exchangeable form [Fig. 3(b)]. Furthermore, studies indicated that the weathering of these particles leads to an increase in Sr-90 mobility in the Chernobyl area. This increase leads to higher Sr-90 concentrations in freshwater and vegetation, resulting in an increase in the Sr-90 dose to populations over time [14].

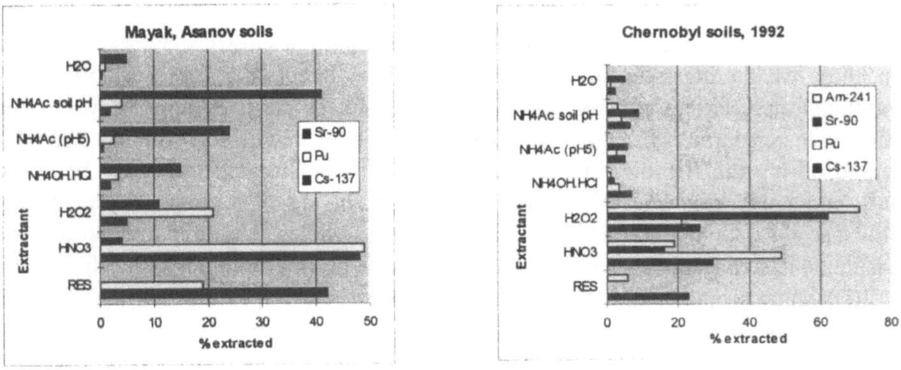


*Figure 3 (a and b).* Sequential extraction of soils and sediments from: (a) Mayak [10,11]; and(b) Chernobyl [13, 15].

## 2.4. STRUCTURE AND COMPOSITION: ELECTRON MICROSCOPY AND SYNCHROTRON RADIATION BASED TECHNIQUES

Scanning electron microscopy (SEM) and transmission electron microscopy (TEM) are powerful techniques for studying the structure of micrometer and sub-micrometer particles. When interfaced with X-ray microanalysis (XRMA), additional information can be obtained on the composition and distribution of elements within the particle surface. The techniques can be applied to the study of suspended sediments and colloids in natural waters and fuel reprocessing effluents [16] and to the study of radionuclide-containing particles in soils and sediments, for example, uranium oxide fuel particles released from Chernobyl or plutonium particles from weapons tests [12,15]. A clear difference in particle structure can be seen in uranium oxide particles released during the initial explosion, collected to the west of the Chernobyl reactor, as compared to those released during the reactor fire collected to the north of the reactor (Fig. 4).

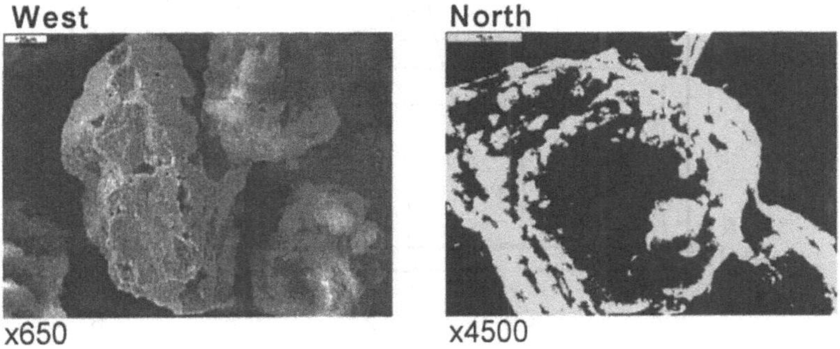


*Figure 4.* A SEM micrograph of Chernobyl particles collected to the west and north of the reactor.

In the Chernobyl exclusion zone, the rates of increase in Sr-90 mobility are determined by the weathering rates of uranium oxide particles in soils and sediments. In turn, these weathering rates are dependent on soil/sediment chemistry (i.e. pH, redox conditions) and particle composition [14]. Deposition of particles released under the initial explosion formed a radioactive trace to the west of the reactor; whereas, particles released during the fire (i.e., under oxidising conditions) formed traces to the north, south, and north east (Fig. 5). Synchrotron radiation based microtechniques enable solid-state speciation to be carried out on single particles. Uranium oxidation state measurements on sections of Chernobyl particles using microscopic X-ray diffraction (u-XRD), microscopic X-ray absorption near-edge structure spectroscopy (u-XANES), confirmed that differences in weathering rates could be correlated with the uranium oxidation state, and showed that the oxidation state of the released reactor fuel particles depended on the release scenario [17]. During the reactor fire, the uranium oxide fuel was oxidised, and particles collected from the northern and southern trace had a $UO_2$ core with surrounding layers of $U_2O_5/U_3O_8$. Particles collected from the Western trace had a $UO_2$ core surrounded by layers of $UO_{2-x}$, suggesting that reduction had taken place under the initial release conditions, probably owing to reactions with the graphite moderator (Fig. 5).

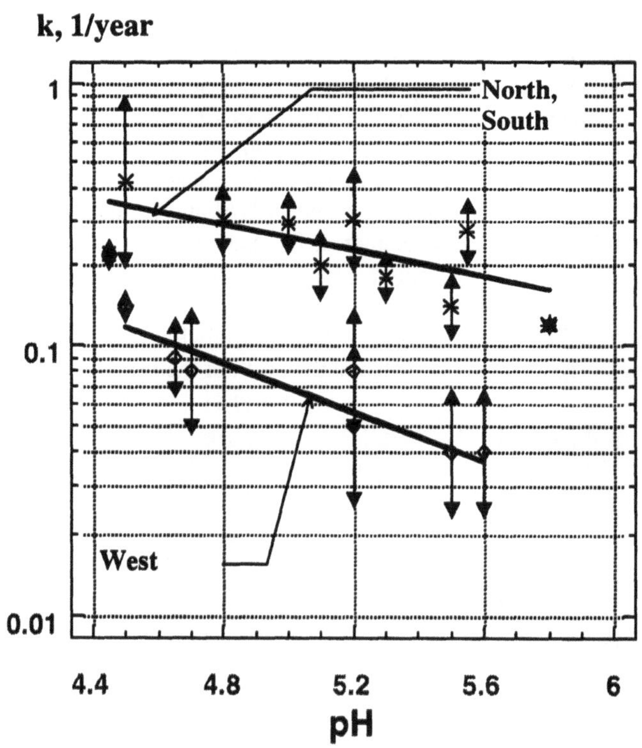


*Figure 5.* Correlation of fuel particle dissolution constants (k, $yr^{-1}$) with soil acidity - $pH_{H2O}$[14].

## 3. Low-Level Measurement Techniques: Mass Spectrometry and Isotope Ratios

For many long-lived radionuclides, mass spectrometric techniques offer the advantage of low detection levels and simpler radiochemical separation as compared to the results of standard counting methods. Recent developments in accelerator mass spectrometry (AMS) of heavy elements have opened the possibility of using the technique for heavy radionuclides such as Pu, Np and Tc [18]. The high sensitivity of AMS enables the measurement of $^{239}Pu$ activities down to levels which are well below those detectable by routine alpha spectrometry. On the basis of procedural blanks, the detection limit for Pu isotopes is less than 1 fg Pu [19,20]. A further advantage of AMS is that it gives information on Pu-isotope ratios that can act as a fingerprint for different sources and act as a tracer in aquatic systems. The ratio is known to vary with reactor type, nuclear fuel burn-up time, neutron flux and energy, fallout from nuclear tests, weapon/device type and yield.

### 3.1. ACCELERATOR MASS SPECTROMETRY

Analytical development work and AMS measurements have been carried out using the 14UD tandem accelerator at the Australian National University, Canberra [18, 19]. The initial chemistry is similar to for alpha spectroscopy: the Pu-isotopes are separated from the other radionuclides by ion exchange chromatography by using high purity $^{242}Pu$ as a yield monitor. After addition of Fe salt, extracts are taken to dryness with nitric acid, then ashed to give final samples for AMS measurement as plutonium dispersed in $Fe_2O_3$ with a nominal $^{242}Pu$:Fe atom ratio or $3x10^{-10}$. The $Fe_2O_3$ samples are mixed with Al powder, pressed into Al holders and mounted on a sample wheel. For isotope measurement, PuO- ions from the negative-ion source are injected into the 14UD accelerator and accelerated to 4.3 MeV. After further acceleration, $^{A}Pu^{6+}$ ions are selected by an analysing magnet and counted individually using a longitudinal-field ionisation detector. The three plutonium isotopes (mass 242, 240 and 239) are counted sequentially using repeat cycles for each sample, and calibrated against a certified Pu-isotope ratio reference standard (Atomic Energy Authority (AEA) Technology).

AMS has been used to determine Pu activities and $^{240}Pu/^{239}Pu$ isotope ratios in samples contaminated by releases from the "Mayak" Production Association. The samples were collected during joint Norwegian-Russian field work in 1994 and 1996, as part of bilateral investigations on sources contributing to radioactive contamination of the Techa River and the surrounding area [10]. Results indicated that the ratio of $^{240}Pu$ up to the present $^{239}Pu$ in waste has increased [20]. The lowest $^{240}Pu/^{239}Pu$ atom ratios, 0.014 - 0.024, were found at the Asanov Swamp, where the primary source of contamination was discharge of radioactive waste to the Techa River between 1949 and 1951. The highest ratios, 0.06 - 0.29, were found in industrial reservoirs contaminated by various sources of waste up to the present day (Fig. 6). Measurement of Pu isotopes in sediment and water samples collected from the Techa, Iset, and Ob rivers showed that while activity levels decrease with distance from Mayak, $^{240}Pu/^{239}Pu$ isotope ratios increase. These results suggest that most of the plutonium in the upper Techa River originates from the early waste discharges. However, enhanced atom ratios in surface and suspended sediments (0.035 - 0.099) indicate an influence from other sources [20].

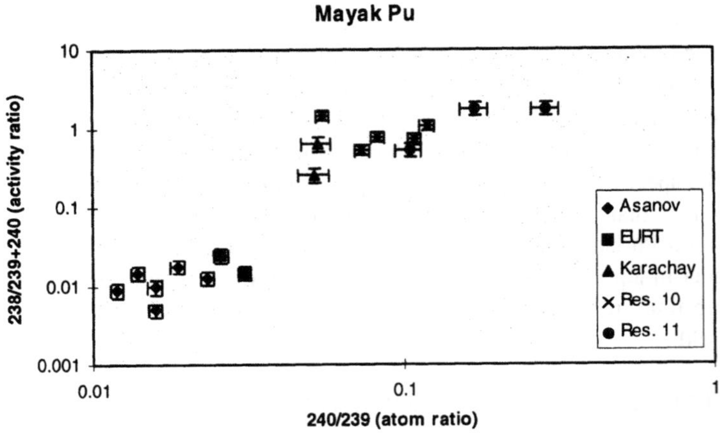


*Figure 6.* Plutonium isotope ratios in samples from Mayak. Dates of main contamination: Asanov, 1949-1951; East Urals Radioactive Trace (EURT), 1957; Karachay, 1951-present; Reservoirs 10 and 11: 1952-present [20].

## References


1. Salbu, B and Oughton, D.H. (1995) Processes affecting the uptake of radioactive species into the environment, *Rad. Prot. Dosimetry*, **62**, 1-4.
2. Salbu, B. and Oughton, D.H. (1994) Analytical strategies for the determination of physico-chemical forms of trace elements in natural waters, in *Trace Elements in Natural Waters* (ed. B.Salbu and E. Steinnes), CRC Press, pp. 41-69.
3. Salbu, B. and Bjørnstad, H.E. (1990) Analytical techniques for determining radionuclides associated with colloids in waters. *J. Radioanal. Nuc. Chem.* **139**, 337-346.
4. Nelson, D.M. and Lovett, M.B. (1978) Oxidation state of plutonium in the Irish Sea *Nature*, **276**, 599-601.
5. Skipperud, L., Oughton, D.H. and Salbu B. Impact of plutonium oxidation state on distribution coefficients in soil-water systems, *Sci. Tot. Environ.* submitted
6. Oughton, D.H., Børretzen, P., Salbu, B. and Trondstad, E. (1997) Mobilization of radionuclides from sediments: potential sources to Arctic Waters, *Sci. Tot. Environ.* **202**, 155-165.
7. Oughton, D.H. and Salbu, B. (1994) Fractionation techniques applicable for trace element speciation in soil-water systems, in *Trace Elements in Soils and Plants: Research Methodology* (ed. H. Bergseth), Agricultural Research Centre of Finland, Jokioinen, pp. 69-77.
8. Oughton, D.H., Salbu, B., Riise, G., Lien, H., Østby, G. and Nøren, A. (1992) Radionuclide mobility and bioavailability in Norwegian and Soviet soils, *Analyst*, **117**, 481-486.
9. Kennedy, V., Sanchez, A., Oughton, D.H. and Rowland, A.P. (1997) Use of single and sequential chemical extractants to assess radionuclide and heavy metal availability from soils for root uptake, *Analyst*, **122**, 89R-100R.
10. Joint Norwegian Russian Expert Group (1997) Sources contributing to radioactive contamination of the Techa River and areas surrounding the "Mayak" Production Association, Urals, Russia.

Joint Norwegian-Russian Expert Group for Investigation of Radioactive Contamination in the Northern Areas: Østerås.

11. Trondstad, E., Nøren, A., Oughton, D.H., Salbu, B., Skipperud, L., Clacher, A., Romanov, G., Strand, P. and Christensen, G.C. (1995) Mobilization of 90Sr and other radionuclides from Mayak and Chernobyl sediments, in *Radioactivity in Arctic Waters*. (eds. P. Strand and A. Cooke), NRPA, Østerås, pp. 256-260.
12. Salbu, B., Krekling, T., Oughton, D.H., Østby, G., Kashparov, V.A., Brand, T.L. and Day, J.P. (1994) Hot particles from accidental releases from Chernobyl and Windscale installations, *Analyst*, **119**, 125-130.
13. Oughton, D.H., Salbu, B., Brand, T.L., Day, J.P. and Aarkrog, A. (1993) Under-determination of strontium-90 in soils containing particles of irradiated uranium oxide fuel, *Analyst*, **118**, 1101-1105.
14. Kashparov, V.A., Oughton, D.H., Protsak, V.P., Zvarisch, S.I., Protsak, V.P. and Levchuk, S.E. (1999) Kinetics of fuel particle weathering and $^{90}$Sr mobility in the Chernobyl 30 km exclusion zone, *Health Physics*, **76**, 251-259.
15. Salbu, B., Krekling, T. and Oughton, D.H. (1998) Analytical methods for studying hot particles, *Analyst*, **123**, 843-849.
16. Salbu, B., Bjørnstad, H.E., Svaren, I., Prosser, S.L., Bulman, R.A., Harvey, B.R. and Lovett, M.B. (1993) Size distribution of radionuclides in nuclear fuel reprocessing liquids after mixing with seawater. *Sci. Tot. Environ.* **130/131**, 51-63.
17. Salbu, B., Janssens, K., Krekling, T., Simionovici, A., Drakopoulous, M., Raven, C., Snigireva, I., Snigirev, A., Linf, O.C., Oughton, D.H. Adams, F. and Kashparov, V.A. Initial release of reduced U particles and subsequent release of oxidised $U_2O_5/U_2O_8$ particles during the Chernobyl reactor accident. *Science*, submitted.
18. Fifield, L.K., Cresswell, R.G., di Tada, M.L., Ophel, T.R., Day, J.P., Clacher, A.P., King, S.J. and Priest, N. (1996) Accelerator mass spectrometry of plutonium isotopes. *Nuc. Instr. Meth. Phys. Res. B.* **117**, 295-303.
19. Oughton, D.H., Fifield, L.K., Day J.P., Cresswell, R.C., Skipperud, L.S. and Salbu, B. (1999) Determination of $^{240}$Pu/$^{239}$Pu isotope ratios in Kara Sea and Novaya Zemlya sediments using accelerator mass spectrometry, in *Proc. Conf. Marine Pollution*, IAEA, Vienna, pp. 123-128 (1999).
20. Oughton, D.H., Fifield, L.K., Day J.P., Cresswell, R.C., Salbu, B. and Skipperud, L.S. Plutonium from Mayak: Measurements using Accelerator Mass Spectrometry, *Environ. Sci. Technol.*, in press.

# RADIATION SURVEY AND RADIONUCLIDE ANALYSIS OF RADIOACTIVE PLUMES FROM THE FORMER SEMIPALATINSK NUCLEAR TEST SITE

O.I. ARTEMEV, M.A. AKHMETOV, V.N. LARIN, L.D. PTITSKAYA,
K.SH. ZHUMADILOV, G.P. SILKINA, S.B. SUBBOTIN
*Institute of Radiation Safety and Ecology*
*National Nuclear Centre*
*Kurchatov-City, Republic of Kazakhstan*

## 1. Background

The radioactive products (RPs) present within the areas of atmospheric nuclear testing in the territory of the Experimental Field and within the areas of radioactive fallout create a complicated radiation situation at the Semipalatinsk Test Site (STS) that has evolved over time. By virtue of their generation, RPs vary from several microns to 10 cm. By virtue of their size, powdery fractions of radioactive products are capable of being resuspended by the wind and transported far away from the place of origin. The areas with a $^{137}$Cs contamination level higher than 1 Ci per km$^2$ occupy about 0.5% of the total STS area. The total $^{137}$Cs activity is estimated at 3000 Ci. The total $^{90}$Sr activity is estimated at less than 2500 Ci [1, 2].

The present radionuclide contamination by $^{239,240}$Pu, $^{241}$Am, $^{90}$Sr, and $^{137}$Cs generally stems from the first above-ground nuclear tests, including the first thermonuclear explosion (12 September 1953) whose plume extends to the southeast from the Experimental Field [2]. The $^{137}$Cs content in the epicentral part of the plume amounts to 5 Ci per km$^2$.

When carrying out fieldwork along the radioactive plume "footprints" at the Experimental Field, we based our objectives on the need to accomplish the following tasks:

- Identify radionuclides present within a 30-km radius from ground zero, their distribution patterns, and the species in soil and slag;
- Determine radionuclide content, distribution patterns, and species (particularly, $^{239,240}$Pu, $^{241}$Am, and $^{90}$Sr) present within southeastern and southwestern radioactive plume "footprints" identified by the aerial gamma survey;
- Study radionuclide spatial and vertical distribution that is necessary for better understanding of migration capabilities.



*S.S. Hecker et al. (eds.), Nuclear Physical Methods in Radioecological Investigations of Nuclear Test Sites*, 127-143.
© 2000 *Kluwer Academic Publishers. Printed in the Netherlands.*

## 2. Surveying Plume Footprints

A nuclear test plume footprints survey that has been performed includes exposure dose rate measurements, determination of α and β contamination levels, soil sampling, and topographic fixing of measurement and sampling points.

Samples for laboratory analysis have been obtained using the following two procedures:

a. Four lateral sections of latitudinal direction have been set up across the southwest plume, the first one being 220 m from the nuclear tests' ground zero (crater edge). Five sampling and measuring points have been chosen at each lateral section every 150–180 m from ground zero. The lateral sections are hereafter referred to as Profiles I, II, III, and IV. Profiles I and II were set up within the slag area, Profile III—in the area of background gamma levels (up to 40 μR per hour), and Profile IV—traversing the spot 1.5 km away from the south border of the Experimental Field (Fig. 1). Samples have been taken from 20 locations—5 locations at every profile. Sampling has been performed to a depth of 20 cm by using special sampling tools and a sampling procedure called the "envelope." Surface slag samples have been taken from Points 1 to 5 of Profiles I and II (altogether 10 slag samples). Soil depth profile samples, every layer being 2 cm thick and 200 $cm^2$ in area, and soil depth profile samples, every layer being 5 cm thick and 100 $cm^2$ in area, have been collected from Points 1, 3, 5 and Points 2, 4, of all profiles, respectively.

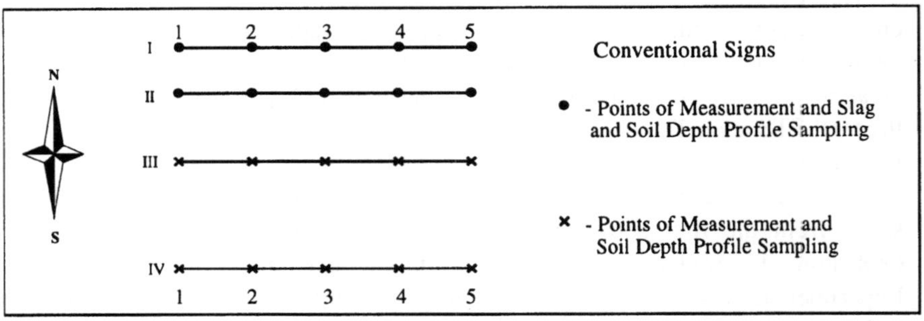


*Figure 1.* Profile and Sampling Point Layout.

b. Eight radial survey routes have been plotted from the epicentre (Fig. 2):
   1. The southeastern (SE) route extended up to the 115th km along the southeast plume footprint;
   2. The southwestern (SW) route extended for 42 km along the southwest plume footprint;
   3. The western (W) route extended for 30 km to the west;
   4. The northwestern (NW) route extended for 30 km to the northwest;
   5. The northern (N) route extended for 30 km to the north;

6. The northeastern (NE) route extended for 30 km to the northeast;
7. The eastern (E) route extended for 30 km to the east;
8. The southern (S) route extended for 30 km to the south.

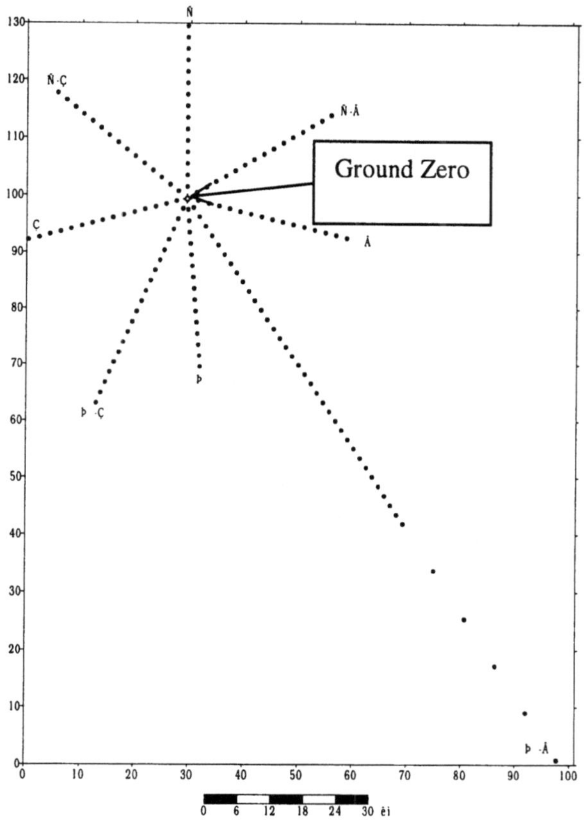


*Figure 2*. Sampling Point Layout.

Within the first 70 km of the SE route, samples were taken every 2 km; within the rest of the route they were taken every 10 km. As for other routes, samples were collected every 2 km away from the epicentre (110 samples). According to the second procedure, soil samples were collected by using a sampling routine called the "triangle"; each side of the triangle was 150 m long. A sample taken using this procedure consisted of three samples, each being 30x30x5 cm, collected from triangle apices and mixed together. The central point of every triangle was fixed by using a satellite navigation instrument.

The data obtained made it possible to determine the characteristic feature of radionuclide distribution over the space and depth of the top soil.

A considerable area occupied by the southwest plume footprint, especially within Profiles I and II (Fig. 1), is covered with slag. Slag samples analysed by gamma spectrometry showed the following radionuclide content measured in Bq per kg: $^{137}Cs$ up to 150 000, $^{241}Am$ over 3000, $^{152}Eu$ up to 130 000, and $^{60}Co$ over 5000. Slag samples have been collected and analysed by the Institute of Radiation Safety and Ecology jointly with the Institute of Nuclear Physics of the National Nuclear Centre of the Republic of Kazakhstan (Tables 1 and 2).

Radionuclide concentrations measured in soil depth profile samples taken from Profile II are presented in Table 1. On the basis of this data, all radionuclides detected can be divided into three groups by their generation:

1. $^{152}Eu$ and $^{60}Co$ (terrigenic activation products)—poor depth-content dependence;
2. $^{241}Am$ (charge substance) and $^{137}Cs$ (charge fission products)—strong depth-content dependence;
3. $^{154}Eu$ (charge fission products + terrigenic activation products)—average depth-content dependence.

TABLE 1. Gamma Spectrometry Results of Soil Samples Taken Near Ground Zero in Bq per kg ± error in %

| Soil Layer (cm) | Specific Content | | | | |
|---|---|---|---|---|---|
| | $^{137}Cs$ | $^{241}Am$ | $^{152}Eu$ | $^{154}Eu$ | $^{60}Co$ |
| 0-2 | 3609±11 | 234±15 | 94483±8 | 4741±15 | 4217±9 |
| 2-4 | 1343±15 | 113±30 | 92083±8 | 4556±15 | 3982±9 |
| 4-6 | 1383±14 | 72±50 | 90028±8 | 3046±15 | 3826±9 |
| 6-8 | 432±18 | 32±69 | 68625±8 | 2370±15 | 3142±9 |
| 0-2 | 46183±8 | 808±10 | 66554±8 | 2838±15 | 3368±9 |
| 2-4 | 32383±8 | 556±12 | 55691±8 | 1818±15 | 2525±9 |
| 4-6 | 30065±8 | 414±20 | 57654±8 | 1988±20 | 2510±9 |
| 0-2 | 6149±9 | 415±15 | 26648±8 | 1540±15 | 1061±12 |
| 2-4 | 1228±11 | 112±20 | 24000±8 | 960±15 | 949±12 |
| 4-6 | 1555±12 | 120±20 | 20227±8 | 679±15 | 800±12 |
| 0-2 | 666±11 | 26±40 | 1164±11 | 97±30 | 59±30 |
| 2-4 | 525±11 | 34±60 | 1022±12 | <5 | 8±100 |
| 4-6 | 602±12 | 64±20 | 925±11 | 56±50 | 33±30 |
| 6-8 | 548±12 | 11±100 | 775±11 | 28±50 | 34±30 |

TABLE 2. Results of Gamma Spectrometry of Soil Samples From the Experimental Field (measured in Bq per kg ± error in %)

| Line # | Sampling Point | Sampling Depth (cm) | Specific Content | | | | |
|---|---|---|---|---|---|---|---|
| | | | Co-40 | Cs-137 | Th-232 | U-238 | Am-241 |
| Profile III | | | | | | | |
| 1. | P 1. L 4 | 6-8 | 432±13 | 13±50 | 21±26 | 20±14 | - |
| 2. | L 5 | 8-10 | 393±15 | 7±90 | 19±17 | 18±45 | - |
| 3. | L 6 | 10-12 | 551±10 | 11±50 | 18±36 | 17±23 | - |
| 4. | L 7 | 12-14 | 594±10 | 23±30 | 1•±15 | 1•±22 | - |
| 5. | L 8 | 14-16 | •14±7 | 42±6 | 21±8 | 24±10 | 10±27 |
| 6. | L 9 | 16-18 | 516±10 | <6 | 22±16 | 16±19 | - |
| 7. | L 10 | 18-20 | 546±10 | 7±90 | 25±27 | 18±37 | - |
| 8. | P 3. L 4 | 6-8 | 583±8 | 19±11 | 20±9 | 24±11 | - |
| 9. | L 5 | 8-10 | •44±8 | 14±14 | 22±9 | 28±10 | - |
| 10. | L • | 10-12 | 667±8 | 13±13 | 20±11 | 27±11 | - |
| 11. | L 7 | 12-14 | 714±7 | 9±20 | 23±8 | 24±13 | - |
| 12. | L 8 | 14-16 | 847±6 | 20±15 | 30±15 | 30±20 | - |
| 13. | L 9 | 16-18 | •80±7 | 10±16 | 22±9 | 27±12 | - |
| 14. | L 10 | 18-20 | 569±9 | 8±22 | 28±8 | 28±12 | - |
| 15. | P 4. L 1 | 0-5 | 512±10 | 203±10 | 18±22 | 17±24 | 203±19 |
| 16. | L 2 | 5-10 | 552±10 | 15±50 | 22±22 | 16±20 | - |
| 17. | L 3 | 10-15 | 200±20 | <6 | 15±27 | 12±26 | - |
| 18. | L 4 | 15-20 | 627±10 | 14±50 | 23±48 | 15±22 | - |
| 19. | P 5. L 1 | 0-2 | 610±12 | 179±7 | 17±21 | 19±27 | 307±9 |
| 20. | L 2 | 2-4 | 760±11 | 63±13 | 27±13 | 31±17 | 100±17 |
| 21. | L 3 | 4-6 | 576±15 | 7±95 | 23±14 | 25±14 | 18±21 |
| 22. | L 4 | 6-8 | 453±20 | 7±100 | 24±14 | 32±17 | 10±100 |
| 23. | L 5 | 8-10 | 673±15 | 7±100 | 23±15 | 26±17 | 10±100 |
| 24. | L 6 | 10-12 | 750±11 | 7±100 | 20±16 | 3215 | 25±58 |
| 25. | L 7 | 12-14 | 572±18 | 7±100 | 19±17 | 20±24 | 10±100 |
| 26. | L 8 | 14-16 | 614±12 | 7±100 | 26±14 | 25±20 | 10±100 |
| 27. | L 9 | 16-18 | 657±8 | 8±20 | 22±8 | 28±14 | - |
| 28. | L 10 | 18-20 | 537±15 | 7±100 | 24±14 | 23±21 | 10±100 |
| Profile IV | | | | | | | |
| 29. | P 1. L 4 | 6-8 | 710±10 | 43±7 | 20±13 | 28±13 | 10±29 |
| 30. | L 5 | 8-10 | 580±13 | 9±78 | 27±12 | 26±17 | - |
| 31. | L 6 | 10-12 | 651±9 | 22±9 | 21±8 | 25±12 | 7±45 |
| 32. | L 7 | 12-14 | 600±12 | 39±20 | 26±14 | 30±17 | - |
| 33. | L 8 | 14-16 | •52±7 | 10±15 | 22±9 | 28±13 | - |
| 34. | L 9 | 16-18 | 5•7±11 | 44±20 | 29±15 | 17±17 | - |
| 35. | L 10 | 18-20 | 516±11 | 2±100 | 22±20 | 16±2 | - |
| 36. | P 3. L 4 | 6-8 | 661±7 | 43±6 | 25±8 | 25±14 | 7±37 |
| 37. | L 5 | 8-10 | 439±20 | 11±30 | 17±30 | 19±35 | - |
| 38. | L 6 | 10-12 | 780±11 | 7±100 | 27±14 | 23±20 | - |

Note: P – sampling point, L – soil layer.

TABLE 2. (Con't)

| Line # | Sampling Point | Sampling Depth (cm) | Specific Content | | | | |
|---|---|---|---|---|---|---|---|
| | | | Co-40 | Cs-137 | Th-232 | U-238 | Am-241 |
| 39. | L 7 | 12-14 | 502±15 | 7±100 | 24±15 | 29±17 | - |
| 40. | L 8 | 14-16 | 711±7 | 71±3 | - | - | - |
| 41. | L 9 | 16-18 | 400±18 | 28±28 | 29±13 | 34±15 | - |
| 42. | L 10 | 18-20 | 478±18 | 7±100 | 25±14 | 37±15 | - |
| 43. | P 4. L 4 | 15-20 | 618±10 | 20±30 | 23±24 | 18±20 | - |
| 44. | P 5. L 4 | 6-8 | 643±7 | 92±3 | 26±9 | 23±17 | 14±30 |
| 45. | L 5 | 8-10 | 780±7 | 60±5 | 25±8 | 30±12 | - |
| 46. | L 6 | 10-12 | 744±7 | 59±5 | 29±8 | 31±10 | - |
| 47. | L 7 | 12-14 | 668±8 | 20±11 | 24±9 | 35±9 | - |
| 48. | L 8 | 14-16 | 743±8 | 18±12 | 27±9 | 30±100 | - |
| 49. | L 9 | 16-18 | •94±8 | 20±10 | 29±10 | 28±14 | - |
| 50. | L 10 | 18-20 | 804±7 | 24±9 | 28±9 | 34±11 | - |

Note: P – sampling point, L – soil layer.

Radionuclide concentrations measured in soil depth profile samples taken from Profiles III and IV are presented in Table 2. Analysis of this data showed that radioactive contamination of soil by anthropogenic radionuclides is uniform and becomes negligible at a 6-cm depth.

As already mentioned, laboratory analysis of samples taken detected the fission products $^{239,240}$Pu, $^{241}$Am, $^{137}$Cs, $^{90}$Sr and activation products $^{152}$Eu, $^{154}$Eu, $^{60}$Co. On the basis of the assumption that $^{137}$Cs is one of the important gamma emitting nuclear fission products that most completely indicates the areas of radioactive fallout, its distribution can be assumed as a basis for determining a radioactive plume "footprint". As for $^{90}$Sr, an assumption was made previously that the farther away from ground zero, the higher the $^{90}$Sr activity and the more likely that a fractionating effect takes place. Therefore, it is necessary to measure $^{137}$Cs and $^{90}$Sr specific contents and their ratios for the same soil samples. Below, a flow diagram of radiological studies carried out by the RK NNC Institute of Radiation Safety and Ecology is shown (Fig. 3).

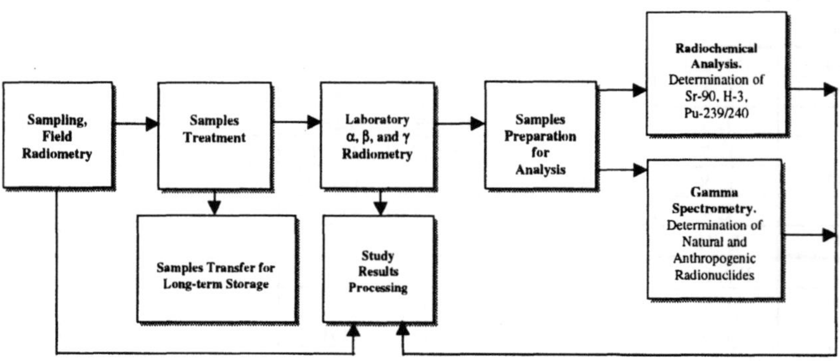


*Figure 3*. Radiological Studies Flow Diagram.

## 3. Measuring Methods

Semiconductive detectors with high-energy resolution connected to the multi-channel pulse analysers are used for determining gamma-emitting radionuclides. Gamma spectra are processed, i.e. radionuclides are identified and quantified by using specialised software installed in a personal computer. Gamma spectrometers are calibrated by using OSGI reference gamma spectral sources.

Standard calibration sources of different geometries are used for precise calibration. The methods used make it possible to detect and quantify low concentrations of 4-6 Bq per kg with a total error of 20% and a confidence level of 95%.

An INSPECTOR field gamma spectral installation based on a GX-1020 HGe detector with the relative efficiency of 10% made by the CANBERRA company is also used for measurements.

In 1999, the Institute was provided with four mobile radiological laboratories equipped with up-to-date field gamma spectrometers as well as dosimetry and satellite navigation instruments.

In 1996–1997 the Institute approved a method of plutonium determination in soil samples used by the Radium Institute Research and Production Association (Saint-Petersburg, Russian Federation). The Institute of Radiation Safety and Ecology (IRSE) began analyses of soil samples for plutonium in 1997 after the radiochemical laboratory was equipped with all the necessary equipment. For two years the Radium Institute's specialists performed selective quality control. Application of this procedure allowed us to determine plutonium concentrations in several hundreds of soil samples taken from different areas of the STS. Currently, in the IRSE's Radioecology Department there is a set of radiochemical methods developed for plutonium determination in soil, plants, and water. The following procedure is common for all radiochemical methods of $^{239,240}Pu$ determination in environmental samples (Fig. 4): sample preliminary treatment; plutonium dissolution; concentrating; plutonium separation from other radionuclides and chemical elements; and separation in the form of a thin-layer sample meeting alpha spectrometry requirements. As plant and water samples treatment for radiochemical analysis does not have particular difficulties and the rest of the procedure is quite similar to that for soil samples, methods for plutonium determination in vegetation and water have been developed and approved in an expeditious manner.

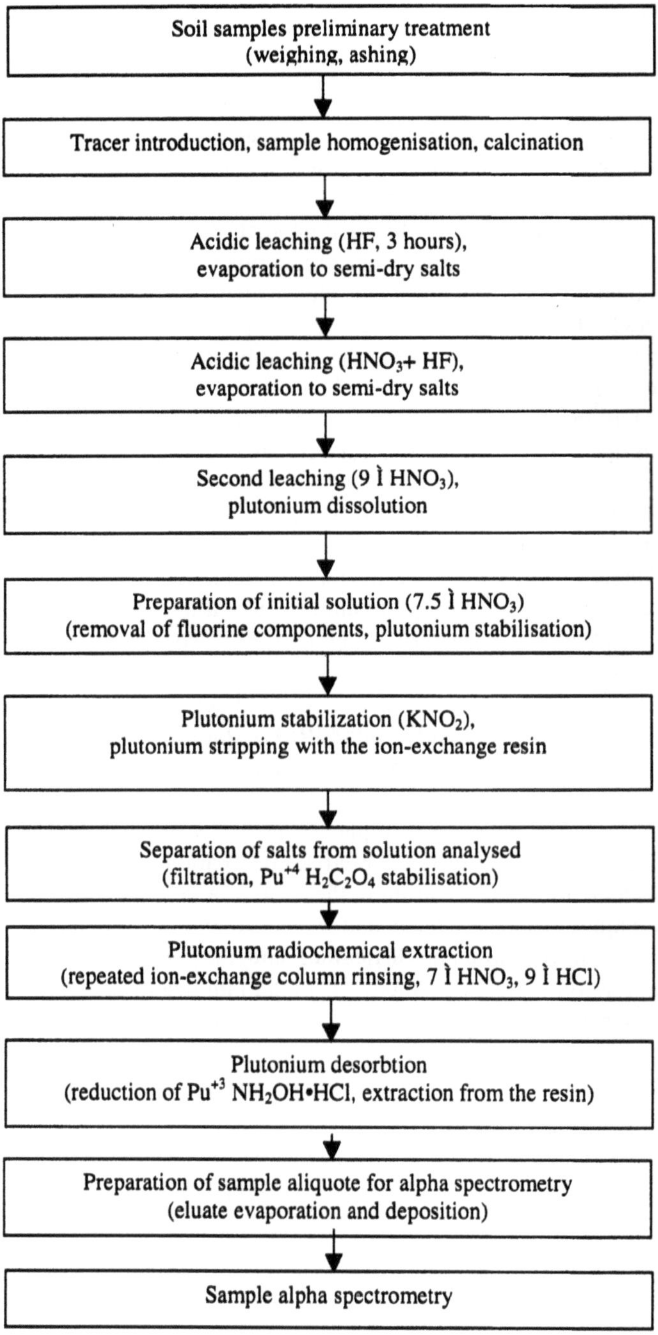


*Figure 4.* Plutonium Determination in a Soil Sample Flow Diagram

Figures 5 and 6 present the gamma and alpha spectra acquired from one of the soil samples taken at the STS.

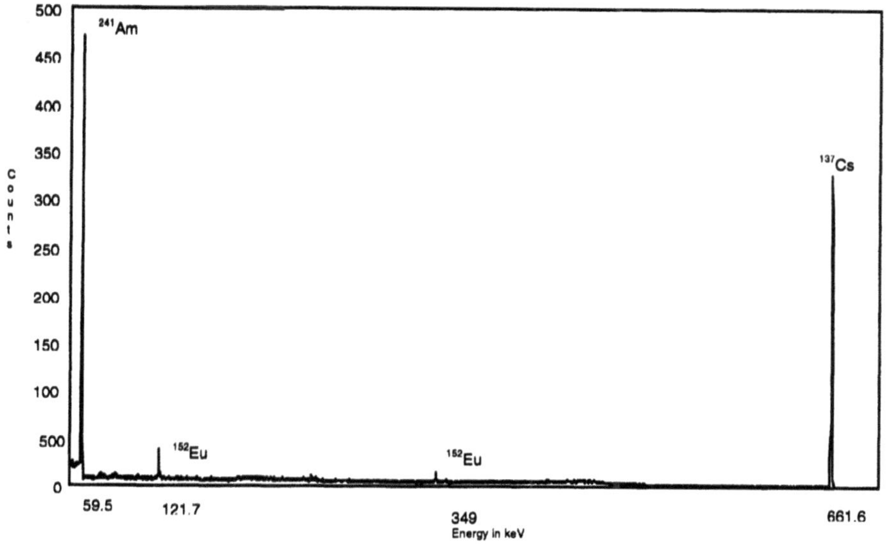


*Figure 5.* Soil Sample Gamma Spectrum

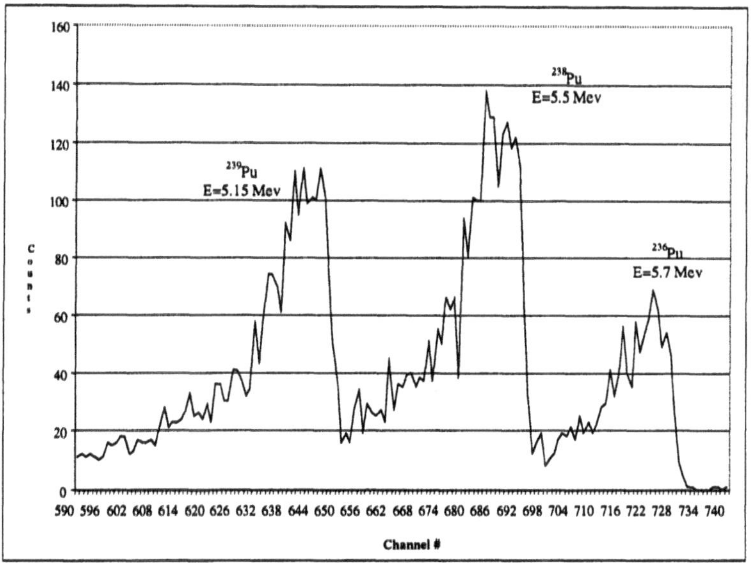


*Figure 6.* Separated Plutonium Alpha Spectrum

## 4. $^{90}$Sr Content Analysis

The same samples have been analysed for $^{90}$Sr content by using the conventional oxalic separation method described below. After separation of chemically pure $^{90}$Sr with the carrier, samples have been set aside for accumulation of daughter $^{90}$Y, whose hard beta radiation was measured with RUB-01P, low-background beta installation.

As already mentioned [3], those methods used for sample destruction with acids that do not involve hydrofluoric acid are intended for determination of $^{90}$Sr water soluble species generally caused by global radioactive fallout. Taking into account data on intractable hot particles obtained previously, it is necessary to determine the total specific activity of $^{90}$Sr in order to correctly interpret the results. A method of the full destruction of samples analysed makes this possible. We used a method of soil sample alkali destruction that is also described in this paper.

Currently, the radiochemical laboratory uses for radiochemical analysis the following measuring installations:

- two-chamber alpha spectrometer (for $^{239,240}$Pu determination);
- RZhS-015 installation (or tritium determination);
- RUB-01P installation (for $^{90}$Sr determination using beta radiation of daughter $^{90}$Y).

### 4.1. $^{90}$Sr OXALIC SEPARATION

This separation process includes the following:

- Twice-repeated sample destruction with 6 M HCl, strontium carrier introduction, filtration with decantation;
- Removal of interfering rare-earth and other radionuclides, iron and manganese dioxides using strong ammonia followed by sodium carbonate, filtration;
- Sample dissolving with HCl and iron removal using non-carbonate ammonia, filtration, acidification, yttrium carrier introduction, 15-day accumulation;
- Yttrium hydroxide precipitation and reprecipitation using non-carbonate ammonia, filtration, rinsing, dissolving;
- Yttrium oxalate precipitation using strong oxalic acid, filtration, rinsing, precipitate drying, and beta radiometry.

### 4.2. $^{90}$Sr ALKALI LEACHING [4]

Radioactive strontium is dissolved using soil sample fusion with alkalis in a porcelain crucible followed by fusion cake leaching. Other radionuclides and many rock-forming elements (silicon, iron, aluminium, manganese, potassium, etc.) are dissolved along with strontium. Therefore, the main task when measuring strontium is its separation from stable elements and the following purification from concomitant radioactive elements. Strontium separation from iron, aluminium, manganese, yttrium, and other interfering elements is carried out using iron and yttrium hydroxides. Further precipitation of the second group's elements makes it possible to clear strontium from caesium traces.

The difference in solubility of crystalline strontium and calcium nitrates in acetone is used for strontium separation from calcium (strontium nitrate is generally insoluble in waterless acetone). After calcium separation, the strontium solution is set aside for 15 days to equilibrate $^{90}$Sr and $^{90}$Y. The yttrium solution is dropped onto a backing and measured with a radiometer. The $^{90}$Sr activity is evaluated by its daughter product $^{90}$Y. The chemical recovery of strontium is determined by titration with B trylon under the presence of zinc and a dark-blue chrome indicator. Chemical recovery of yttrium is determined by gravimetric titration. This method makes it possible to determine $^{90}$Sr specific (cubic) activity with a relative error of less than 0.20 (0.95 confidence level).

Table 3 presents $^{90}$Sr and $^{137}$Cs specific activities and Sr/Cs ratios calculated for points of measurements. The $^{90}$Sr data were obtained by using two methods. One was intended for determining total $^{90}$Sr content including hot particles; the other method (oxalic leaching) was only intended for readily soluble strontium species usually caused by global fallout.

TABLE 3. $^{137}$Cs and $^{90}$Sr Specific Activities and $^{90}$Sr/$^{137}$Cs Ratios Measured within the Southeastern Plume "Footprint."

| Distance from Epicentre (km) | Cs Specific Activity (Bq/kg) | Oxalic Separation | | Alkali Leaching | |
|---|---|---|---|---|---|
| | | $^{90}$Sr Specific Activity (Bq/kg) | $^{90}$Sr/$^{137}$Cs | $^{90}$Sr Specific Activity (Bq/kg) | $^{90}$Sr/$^{137}$Cs |
| 3 | 37 | 32 | 0,86 | 151 | 4,08 |
| 7 | 7 | - | - | 107 | 15,29 |
| 9 | 54 | 50 | 0,93 | 217 | 4,02 |
| 15 | 34 | 12 | 0,35 | - | - |
| 17 | 36 | 76 | 2,11 | 295 | 8,19 |
| 21 | 7 | 74 | 10,57 | 252 | 36,00 |
| 23 | 37 | 152 | 4,11 | 1200 | 32,43 |
| 26 | 36 | 60 | 1,67 | 402 | 11,17 |
| 28 | 344 | 43 | 0,13 | 283 | 0,82 |
| 55 | 339 | 128 | 0,38 | 854 | 2,52 |
| 57 | 468 | 174 | 0,37 | 706 | 1,51 |
| 73 | 97 | 42 | 0,43 | 200 | 2,06 |
| 75 | 101 | 35 | 0,35 | 159 | 1,57 |

All the data obtained on the specific activities of anthropogenic radionuclides are presented in Table 4. (Data on $^{90}$Sr were obtained using the conventional oxalic separation method.)

TABLE 4. Radionuclide Content of the Soil Samples

| Sample | Distance from Epicentre (km) | Specific Activity (Bq/kg) | | | |
|---|---|---|---|---|---|
| | | $^{90}Sr$ | $^{239/240}Pu$ | $^{137}Cs$ | $^{241}Am$ |
| N-1 | 2 | 115 | 368 | 179 | 97 |
| N-2 | 4 | 18 | 57 | 130 | |
| N-3 | 6 | 23 | 81 | 15 | |
| N-4 | 8 | 5 | 9 | 9 | |
| N-5 | 10 | 30 | 62 | 15 | |
| N-6 | 12 | 450 | 58 | 11 | |
| N-7 | 14 | 44 | 12 | 6 | |
| N-8 | 16 | 48 | 190 | 10 | |
| N-9 | 18 | 24 | 9 | 15 | |
| N-10 | 20 | 4 | 74 | 8 | |
| N-11 | 22 | 11 | 8 | 17 | |
| N-12 | 24 | 51 | 23 | 16 | |
| N-13 | 26 | 32 | 40 | 7 | |
| N-14 | 28 | 38 | 59 | 22 | |
| N-15 | 30 | 8 | 93 | 8 | |
| NE-1 | 2 | 269 | 260 | 271 | |
| NE-2 | 4 | 110 | 196 | 116 | |
| NE-3 | 6 | 56 | 250 | 85 | 27 |
| NE-4 | 8 | 47 | 200 | 36 | |
| NE-5 | 10 | 45 | 45 | 63 | |
| NE-6 | 12 | 57 | 26600 | 6 | |
| NE-7 | 14 | 7 | 30 | 74 | |
| NE-8 | 16 | 47 | 200 | 35 | |
| NE-9 | 18 | 97 | 4540 | 19 | |
| NE-10 | 20 | 93 | 1640 | 82 | |
| NE-11 | 22 | 28 | 240 | 34 | |
| NE-12 | 24 | 58 | 280 | 27 | |
| NE-13 | 26 | <3 | 240 | 6 | |
| NE-14 | 28 | 6 | 180 | 11 | |
| NE-15 | 30 | 12 | 240 | 6 | |
| E-1 | 2 | 133 | 4100 | 135 | |
| E-2 | 4 | 58 | 180 | 81 | 11 |
| E-3 | 6 | 93 | 680 | 95 | 10 |
| E-4 | 8 | 47 | 100 | 47 | |
| E-5 | 10 | 42 | 5100 | 61 | 143 |
| E-6 | 12 | 25 | 240 | 36 | 37 |
| E-7 | 14 | 26 | 34 | 18 | |
| E-8 | 16 | 21 | 50 | 29 | |
| E-9 | 18 | 61 | 470 | 75 | |
| E-10 | 20 | 25 | 78 | 43 | 33 |
| E-11 | 22 | 36 | 100 | 36 | |
| E-12 | 24 | 28 | 790 | 44 | |
| E-13 | 26 | 27 | 170 | 62 | 10 |
| E-14 | 28 | 25 | 120 | 101 | |
| E-15 | 30 | 56 | 910 | 68 | |

TABLE 4 (Con't)

| Sample | Distance from Epicentre (km) | Specific Activity (Bq/kg) | | | |
|---|---|---|---|---|---|
| | | $^{90}Sr$ | $^{239/240}Pu$ | $^{137}Cs$ | $^{241}Am$ |
| SE-1 | 0.77 | | | 36 | |
| SE-2 | 3.05 | 32 | | 57 | |
| SE-3 | 5.07 | | | 64 | |
| SE-4 | 6.92 | | | 35 | |
| SE-5 | 9.02 | | | 47 | |
| SE-6 | 10.66 | | | 67 | |
| SE-7 | 12.86 | | | 101 | |
| SE-8 | 15.04 | | 17 | 97 | |
| SE-9 | 17.00 | 75 | | 133 | |
| SE-10 | 19.12 | | | 280 | |
| SE-11 | 21.02 | 169 | 36 | 236 | |
| SE-12 | 23.35 | | 58 | 402 | |
| SE-13 | 26.16 | 217 | 160 | 585 | |
| SE-14 | 28.14 | 299 | | 484 | |
| SE-15 | 30.14 | . | | 468 | |
| SE-16 | 32.15 | 203 | | 339 | |
| SE-17 | 34.40 | | | 262 | |
| SE-18 | 36.58 | 183 | | 309 | |
| SE-19 | 39.11 | | | 551 | |
| SE-20 | 41.05 | | 96 | 568 | |
| SE-21 | 43.11 | | | 393 | |
| SE-22 | 45.11 | 389 | 207 | 572 | |
| SE-23 | 47.36 | | 76 | 415 | |
| SE-24 | 49.66 | | | 276 | 10 |
| SE-25 | 51.53 | 78 | 140 | 132 | 10 |
| SE-26 | 53.53 | | | 176 | 10 |
| SE-27 | 55.24 | 394 | 270 | 349 | 10 |
| SE-28 | 57.31 | | | 185 | 10 |
| SE-29 | 59.37 | | | 344 | 10 |
| SE-30 | 61.72 | | | 36 | 64 |
| SE-31 | 65.20 | 34 | | 37 | 10 |
| SE-32 | 66.73 | | | 7 | 10 |
| SE-33 | 68.73 | | | 69 | 10 |
| SE-34 | 70.78 | | | 36 | 10 |
| SE-35 | 72.76 | | | 34 | 46 |
| SE-36 | 74.77 | | | 48 | 10 |
| SE-37 | 82.71 | 70 | | 96 | 427 |

TABLE 4 (Con't)

| Sample | Distance from Epicentre (km) | Specific Activity (Bq/kg) | | | |
|---|---|---|---|---|---|
| | | $^{90}Sr$ | $^{239/240}Pu$ | $^{137}Cs$ | $^{241}Am$ |
| SE-38 | 90.85 | | | 54 | 10 |
| SE-39 | 98.46 | | | 7 | 10 |
| SE-40 | 106.07 | 14 | | 7 | 10 |
| SE-41 | 112.62 | | | 37 | 10 |
| SE-42 | 119.71 | | | 313 | 96 |
| S-1 | 2 | 23 | 140 | 100 | 10 |
| S-2 | 4 | 3S | 80 | 59 | 10 |
| S-3 | 6 | 28 | 200 | 30 | 10 |
| S-4 | 8 | 52 | 128000 | 82 | 11112 |
| S-5 | 10 | 154 | 195 | 129 | 193 |
| S-6 | 12 | 57 | 2300 | 87 | 1452 |
| S-7 | 14 | 11 | 1570 | 23 | 50 |
| S-8 | 16 | 15 | 1100 | 24 | 10 |
| S-9 | 18 | 18 | 370 | 19 | 29 |
| S-10 | 20 | 14 | 260 | 13 | 10 |
| S-11 | 22 | 19 | 98 | 28 | 10 |
| S-12 | 24 | 17 | 480 | 9 | 10 |
| S-13 | 26 | 20 | 110 | 13 | 10 |
| S-14 | 28 | 31 | 100 | 25 | 10 |
| S-15 | 30 | 34 | 160 | 8 | 10 |
| SW-1 | 2 | 137 | 460 | 45 | 10 |
| SW-2 | 4 | 224 | 60 | 288 | 10 |
| SW-3 | 6 | 316 | 250 | 524 | 76 |
| SW-4 | | 542 | 6400 | 775 | 237 |
| SW-5 | 10 | 168 | 500 | 1310 | 220 |
| SW-6 | 12 | 390 | 330 | 596 | 95 |
| SW-7 | 14 | 124 | 105 | 179 | 32 |
| SW-8 | 16 | 300 | 2200 | 695 | 10 |
| SW-9 | 18 | 23 | 140 | 356 | 10 |
| SW-10 | 20 | 360 | 560 | 312 | 10 |
| SW-11 | 22 | 68 | 1200 | 671 | 29 |
| SW-12 | 24 | 33 | 2700 | 890 | 10 |
| SW-13 | 26 | 341 | 409 | 569 | 10 |
| SW-14 | 28 | 43 | 430 | 489 | 34 |
| SW-15 | 30 | 158 | 110 | 276 | 29 |
| SW-16 | 32 | 103 | 206 | 333 | 23 |
| SW-17 | 34 | 100 | 6700 | 232 | 13 |
| SW-18 | 36 | 52 | 170 | 126 | 10 |
| SW-19 | 38 | 371 | 120 | 565 | 43 |
| SW-20 | 40 | 134 | 145 | 204 | 41 |
| W-1 | 2 | 97 | 340 | 137 | |
| W-2 | 4 | 78 | 610 | 114 | 63 |
| W-3 | 6 | 17 | 165 | 18 | |

TABLE 4 (Con't)

| Sample | Distance from Epicentre (km) | Specific Activity (Bq/kg) | | | |
|---|---|---|---|---|---|
| | | $^{90}Sr$ | $^{239/240}Pu$ | $^{137}Cs$ | $^{241}Am$ |
| W-4 | 8 | 14 | 89 | 6 | |
| W-5 | 10 | 19 | 133 | 25 | |
| W-6 | 12 | 11 | 300 | 19 | |
| W-7 | 14 | 17 | 50 | 15 | |
| W-8 | 16 | 11 | 580 | 15 | |
| W-9 | 18 | 27 | 500 | 26 | |
| W-10 | 20 | 53 | 810 | 37 | |
| W-11 | 22 | 35 | 85 | 30 | |
| W-12 | 24 | 30 | 90 | 34 | |
| W-13 | 26 | 18 | 87 | 21 | |
| W-14 | 28 | 48 | 130 | 40 | |
| W-15 | 30 | 29 | 14 | 9 | |
| NW-1 | 2 | 78 | 150 | 79 | 180 |
| NW-2 | 4 | 14 | 47 | 23 | |
| NW-3 | 6 | 14 | 1290 | 17 | |
| NW-4 | 8 | 12 | 1790 | 7 | |
| NW-5 | 10 | 17 | 220 | 15 | |
| NW-6 | 12 | 37 | 315 | 19 | 31 |
| NW-7 | 14 | 22 | 132 | 17 | |
| NW-8 | 16 | 106 | 150 | 20 | |
| NW-9 | 18 | 33 | 280 | 20 | |
| NW-10 | 20 | 8 | 140 | 9 | |
| NW-11 | 22 | 42 | 410 | 12 | |
| NW-12 | 24 | 23 | 260 | 27 | |
| NW-13 | 26 | | 85 | 7 | |
| NW-14 | 28 | 25 | 520 | 23 | |
| NW-15 | 30 | 85 | 50 | 30 | |

## 5. Data Findings

Data in Table 4 confirm the presence of radioactive "spots" in the territory of the test site. Relatively "clean" locations alternate with highly contaminated ones. It should also be noted that, as a whole, the radioactive contamination becomes apparent from 4–6 km away from the epicentral zone. This contamination is obviously related to a great scope of post-test works carried out within the epicentre.

In addition, several peculiarities of contamination by individual radionuclides are to be noted as follows:

- A very high $^{90}Sr$ level as compared to that of $^{137}Cs$ was measured in Point N-6 of the relatively "clean" north route—450 Bq/kg for $^{90}Sr$ and 11 Bq/kg for $^{137}Cs$;
- Laboratory analysis (gamma spectrometry and radiochemistry) of the soil sample from Point E-5 showed a significant $^{239,240}Pu$ contamination level (5100 Bq/kg) and negligible $^{90}Sr$ and $^{137}Cs$ contamination levels. A repeated

radiological survey was made at this point by using field gamma spectrometry. Data on $^{241}$Am (plutonium daughter product) obtained as a result of the second survey did not confirm a high plutonium content;

- Another example of no correlation among contaminants: $^{239,240}$Pu was found in Sample S-4 (128 000 Bq/kg) while Sr and Cs concentrations were found to be negligible.

Some of the cases observed allow us to assume that the radioactive contamination of some areas is rather associated with other experiments carried out at the Experimental Field than above-ground nuclear tests.

The highest $^{239,240}$Pu concentrations were observed along southern, southeastern and southwestern routes. It makes it possible to conclude that a significant area is contaminated at least by plutonium. Almost all the routes showed high radioactivity levels of individual radionuclides. It is particularly characteristic for plutonium contamination. The fact that some cases do not allow for reproducibility of plutonium quantities is undoubtedly the evidence of hot particles.

As the contamination nature is extremely different even along the axes of possible fallout plumes, the ground-based (route) survey only allowed a preliminary understanding of radionuclide distribution patterns at the Experimental Field and along the plume footprints. A detailed, ground-based radiation survey of the territory using a dense survey and sampling grid is necessary for revealing and mapping dangerous, highly contaminated STS areas.

Using both methods of $^{90}$Sr determination [5], researchers made several conclusions about $^{239,240}$Pu, $^{241}$Am, $^{90}$Sr, and $^{137}$Cs contamination distribution along the plume footprints of above-ground nuclear tests conducted in Semipalatinsk region. These conclusions are based on $^{239,240}$Pu, $^{241}$Am, $^{90}$Sr, and $^{137}$Cs specific activities and on $^{90}$Sr/$^{137}$Cs ratios measured in soil samples taken from the southeast (thermonuclear) footprint.

## 6. Conclusions

- Southeastern and southwestern plume footprints are the most contaminated.
- The amount of radioactive fallout after the thermonuclear test that falls to the share of $^{90}$Sr water soluble species is several times less than the total content.
- The highest radioactive contamination by $^{239,240}$Pu, $^{241}$Am, $^{90}$Sr, and $^{137}$Cs is often many km away from the ground zero along the plume "footprints."
- The determined Sr/Cs ratios do not agree with those calculated by using the chain fission model.

## 7. References


1. Smagulov S.G., Artemev O.I., Larin V.N., et al. Radiological Situation Assessment at the Semipalatinsk Test Site and in the Adjacent Regions. // Quarterly and annual reports on the Polygon sub-program. – Kurchatov. – 1996-1997.

2. Dubasov Yu.V., Zelentsov S.L., Krasilov G.A., et al. Chronology of Atmospheric Nuclear Tests Conducted at the Semipalatinsk Test Site and Their Radiation Characteristics. // Semipalatinsk Test Site – Altai Research Program's Bulletin. – 1994. - No 4.
3. Artemev O.I., Yashkarova M.G., Orazshanova L.I., et al. Radiochemical Separation of Strontium-90 From Soil Samples Collected at the Semipalatinsk Test Site. // Semei University Bulletin. – 1998. - No 2.- P. 52-57.
4. A Method for Determination of Sr-90 Specific Content in Environmental and Foodstuff Samples Using Radiochemical Separation. AK.65000.00.541MU // Institute of Atomic Energy. - Kurchatov. –1995.
5. Artemev O.I., Zhumadilov K.Sh., Larin V.N., et al. Radiochemical Separation of Strontium-90 From Soil Samples Collected at the Semipalatinsk Test Site. // Semei University Bulletin. – 1998. - No 3-4.

DETERMINATION OF $^{89}$SR AND $^{90}$SR FROM LOW- AND INTERMEDIATE-LEVEL RADIOACTIVE LIQUID WASTE

Á. VINCZE, J. SOLYMOSI
*Miklos Zrinyi National Defence University,*
*Department of Environmental and NBC Security,*
*H-1581 Budapest P.O.B.: 15., Hungary*
*Email: vincze@zmne.hu*

P. ORMAI
*Public Agency for Radioactive Waste Management (PURAM),*
*H-2040 Budaörs, Puskas Tivadar str. 11, Hungary*

G. VOLENT
*Paks Nuclear Power Plant Ltd., Paks,*
*P.O. Box 71, H–7031 Hungary*

**Abstract**

This paper reports on the method, results, and experiences of the $^{89}$Sr and $^{90}$Sr analysis performed on low-and intermediate-level radioactive liquid waste samples from the Paks Nuclear Power Plant. The analysis is based on a combined separation method, which results in the separation and detection of radiostrontium ($^{89}$Sr and $^{90}$Sr) and of $^{90}$Y. After the suitable pretreatment of the samples, radiostrontium is concentrated using the Sr-Spec® (EIChroM Industries, Darien, IL, 60561 USA) chromatographic column. The eluted radiostrontium is precipitated in a thin layer and its activity measured by a low-background, gamma-ray-compensated, alpha-beta detector. The chemical yield is determined by atomic emission spectroscopy. The $^{90}$Y activity of the samples is also measured by the procedure described elsewhere. This measurement is based on the selective adsorbing property of the yttrium-ammonium-oxalate and the yttrium-oxalate double salt mixture of yttrium. Under the proper conditions ( pH, oxalate anion activity, mass ratio of the salts) the quantitative yield of $^{90}$Y can be achieved, as determined by using the $^{88}$Y inner standard. As a result, the radioactive concentration of $^{90}$Sr in the samples could be calculated. It was concluded that $^{89}$Sr was not present in the samples because of the relatively long (in years) storage life of these samples. However, this combined method can be easily adapted to matrices sampled at nuclear test sites.



*S.S. Hecker et al. (eds.), Nuclear Physical Methods in Radioecological Investigations of Nuclear Test Sites,* 145-150.
© 2000 *Kluwer Academic Publishers. Printed in the Netherlands.*

## 1. Introduction

The $^{235}U$ fission process results in a great cumulative yield of strontium isotopes, among which $^{89}Sr$ and $^{90}Sr$ are of special interest. Regarding the biological hazard of these isotopes, the determination of the amount of these isotopes in radioactive waste samples is inevitable from both radiation protection and nuclear environmental protection points of view.

Owing to the continuous character of the beta spectra, selective separation methods are required for determining the activity concentration of the individual nuclides. In recent years the application of solid-phase extraction techniques has been widespread. One of the frequently used materials for the separation of strontium is the Sr-Spec® chromatographic material (EIChroM Industries, Darien, IL, 60561 USA), which was developed by E. P. Horwitz and co-workers [1-6].

This strontium selectivity is based on the 1-octanol solution of a diciclohexano-18-Crown-6 (DC18C6) type crown ether impregnated on AMBERLIT-XAD7 resin.

The determination of $^{90}Sr$ in samples of nuclear power plant (NPP) origin is usually based on the selective separation of strontium followed by beta counting [1-6]. If the secular radiochemical equilibrium is already established in the sample, the analysis can also be performed based on yttrium separation and measurement of $^{90}Y$ [7-11]. When the sample is relatively young, $^{89}Sr$ may also be present in the sample; therefore more complex analysis is required [12-14].

In this paper, we present a combined separation method, which results in the separation and detection of radiostrontium ($^{89}Sr$ and $^{90}Sr$) and of $^{90}Y$. Following our procedure, radiostrontium is separated by the Sr-Spec® technology, while $^{90}Y$ is simultaneously separated and determined by a procedure developed in our laboratory, the Radiochemical Laboratory of the Technical University of Budapest.

The usefulness of this complex analysis method has been demonstrated by the analysis of low- and intermediate-level liquid waste (LILW) samples originating from NPP Paks, Hungary. From the analytical point of view, these samples are considered to be very complex matrices since they usually contain several fission as well as activation products, together with a variety of chemicals used in the water chemistry and decontamination procedures of the VVER-440 type reactors. The authors believe that this method can be easily adapted to environmental matrices originating from nuclear test sites.

## 2. Experimental Methods

### 2.1. REAGENTS AND EQUIPMENT

The pH was measured with an OP-208/1 digital pH meter (Radelkis, Hungary, Budapest) equipped with a combined glass electrode. Preparation of the samples was done on a cellulose-based membrane filtre (50 mm in diameter; porosity: 0.45 μm;

Sartorius, AG, Göttingen, Germany). The following solutions were prepared from analytical grade chemicals (Fluka puriss.) and double-distilled water (ddw): saturated oxalic acid, 5 m/m % ammonium-oxalate, 8 M NaOH, 8M $HNO_3$, 1M $Na_2S$, 0.1M $NaCO_3$, $Co(NO_3)_2$ $c_{Me}$=2mg/ml, $CsNO_3$ $c_{Me}$=2mg/ml, $YCl_3$ $c_{Me}$=2mg/ml, $SrCl_2$ $c_{Me}$=2mg/ml, $AgNO_3$ $c_{Me}$=2mg/ml. For defecation, Chemviron activated carbon BL pulverised activated carbon was used. For the yttrium chemical-yield determination, the solution of $^{88}YCl_3$ (Analytics, Atlanta, Georgia 30318, USA ) was added to the original samples as an inner standard.

## 2.2. SAMPLE PRECONDITIONING

Each liquid sample (V=100 ml) was diluted with 200 ml ddw; the pH was adjusted to 1-2 with 8M $HNO_3$ solution; then 10-10 mg $Sr^{2+}$, $Y^{3+}$, $Cs^{+}$, $Co^{2+}$ and $Mn^{2+}$ carriers were added. This solution was spiked with 20 Bq $^{88}YCl_3$ solution as an inner standard, after which 1g of activated carbon was also added. This solution was boiled for 20 minutes and let to stand for at least 12 hours. This mixture was then filtred (solution I.). The filtrate was ashed at 800°C, then dissolved in 8M $HNO_3$, and finally this solution was added to solution I. To this combined solution 15-20 g oxalic acid were added, with the pH set between 9-10 using 8M NaOH, and the solution was boiled for 15 minutes. After cooling, the precipitate was vacuum filtred and washed two times with 100 ml distilled water; after adding 1 cc $HNO_3$, the precipitate was heated to dryness several times, then dissolved in 50 ml 3M $HNO_3$. This solution was then diluted, and 10 mg $Ag^{+}$ carrier and 5 ml 0.1M $Na_2S$ solution were added. After heating and cooling, the precipitate wased and washed with 50 ml ddw, and the resulting solution was evaporated to 20-30 ml (solution II).

## 2.3. STRONTIUM PREPARATION

The chromatographic column (Sr•Spec®) was preconditioned by being passed through 10 ml of 3M $HNO_3$ solution. Following this procedure, solution II was passed through with the flow velocity of 0.5 ml/min using a peristaltic pump. (The real-time of the last drop coming through the column was recorded - $t_0$.) The column was then rinsed with 20 ml of 3M $HNO_3$ + 0.05M oxalic acid solution. All elutes were then combined and put aside for later use (solution III). The strontium retained on the column was eluted with 30 ml ddw. The pH of the elute was set to 9-10 with $NH_4OH$, and, after heating, 5 ml 0.1M $Na_2CO_3$ were added. After cooling, the precipitate was filtred to a membrane filtre and dried for 30 minutes at 40°C. The chemical yield of the strontium preparation was determined by atomic emission spectroscopy (AES). The chemical purity of the preparation was checked with the isotope excited, X-ray fluorescence technique.

## 2.4. YTTRIUM PREPARATION

Solution III was diluted with 500 ml ddw; 5 ml of saturated oxalic acid solution was added and the pH was adjusted to 1.4-1.5 during mixing by a magnetic stirrer. The mixture of yttrium-ammonium-oxalate and yttrium-oxalate salt prepared as described [7,8] was added to the solution, after which it was stirred for 20 minutes. The solution was then vacuumed through a membrane filtre, rinsed, and then dried at 40°C. The chemical yield was determined by measuring the $^{88}Y$ activity in the preparation, using HPGe gamma spectroscopy.

## 2.5. BETA COUNTING

The samples were counted in a gamma-background-compensated, low-level alpha-beta monitor (HARSHAW-TASC-12) having the following efficiencies: $\mu_{Sr90} = 0.17$, $\mu_{Sr89} = 0.18$ and $\mu_{Y90} = 0.19$ for $^{90}Sr$, $^{89}Sr$, and $^{90}Y$ respectively. The typical background intensity was 1-2 cpm, while for the sample containing only the $^{88}Y$ inner standard, it was 3.2 cpm. The effect of $^{88}Y$ on the background intensity (due to conversion electrons and positrons) was taken into account by calculating the chemical yield.

The total activity of the strontium preparation extrapolated to $t_o$ time is:

$$A_t^0 = A^{90}Sr + A^{89}Sr \quad (1)$$

At $t_1 > t_0$ the measured intensity is:

$$I_t^1 = \mu^{90}SrA^{90}Sr[e^{-\lambda_1(t_1-t_0)}] + \mu^{90}YA^{90}Sr\frac{\lambda_2}{\lambda_2-\lambda_1}[e^{-\lambda_1 t_1} - e^{-\lambda_2 t_1}] + \mu^{89}SrA^{89}Sr[e^{-\lambda_3(t_1-t_0)}] + I_b \quad (2)$$

where $I_b$ is the background intensity. By counting the sample at two significantly different times, $A^{89}Sr$ and $A^{90}Sr$ can be determined by applying Eq. (2) twice [14].

The $^{90}Y$ activity of the yttrium preparation is determined by measuring the decay curve of the sample and extrapolating to time $t_o$. In this way, we determine the $^{90}Sr$ activity concentrations of the original sample twice, an effort that increases the probability of a successful analysis. If these values agree within the range of experimental error, the mean of the activity concentrations is accepted. Otherwise, the result of the analysis should be rejected and the whole procedure must be repeated.

## 2.6. MINIMAL DETECTABLE ACTIVITY (MDA)

The calculation of the minimal detectable activity (MDA) of the method is shown for the case of $^{90}Sr$ isotope in Equation (3). Here the background intensity was taken to be 3.2 cpm; the time of the measurement was 60 minutes and the chemical yield was taken to be 70%.

$$MDA = \frac{3.29 * \sqrt{\frac{I_b}{t} + \frac{S_b^2}{n}}}{\mu^{90*}Sr\,\eta^{90*}Sr\,V * 60} = 1.1 Bq / dm^3 \quad (3)$$

where: $V$ = volume of the sample (0.100 $dm^3$),
$I_b$ = background intensity (cpm),
$t$ = measurement time (min),
$S_b$ = standard deviation of background measurements (0.1 cpm),
$n$ = number of background measurements (5).

The MDA values for the other isotopes can similarly be calculated using Equation 3.

## 3. Results

Approximately 250 $m^3$ of LILW are generated annually at the Paks site mainly from the water chemistry and decontamination procedures applied at the power plant. This liquid waste is stored in containers. Samples were collected from the top, the middle, and the bottom of each wastewater liquid container stored at NPP Paks. The total gamma activity of these samples varied between $10^3$ and $10^6$ Bq/kg. Results for the 02TW30B002-coded container are presented in Table 1.

Table 1. Analysis results for the container 02TW30B002 containing 381 $m^3$ evaporation concentrate

| Sample | Sr chemical yield (%) | $^{90}Sr$ (Bq/$dm^3$) |
|---|---|---|
| Top | 65 | 71±8 |
| Middle | 57 | 173 ±15 |
| Bottom | 62 | 414± 35 |
| Sample | Y chemical yield (%) | $^{90}Y$ (Bq/$dm^3$) |
| Top | 67 | 79± 9 |
| Middle | 52 | 165 ± 14 |
| Bottom | 51 | 440± 33 |

## 4. Conclusions

It can be seen from Table 1 that the $^{90}Y$ and $^{90}Sr$ activity concentrations of the original sample agree, which means that the results are trusted. If it is not so, the analysis must be repeated. In practically all cases, the $^{89}Sr$ activity concentration was found to be negligible because the age of the samples (many years).

In most cases, the activity concentrations at the top of the containers were lower then those at the bottom. This difference is probably a result of these containers being kept at a basic pH range. In addition, because some slow precipitation of the ingredients might occur, these ingredients then become concentrated at the bottom of the container due to sedimentation.

## 5. References

1. Horwitz E. P., Dietz M. L. and Fisher D. E.: *Anal. Chem.*, **63**, No. 5, 1991

2. Vajda N., Ghods-Esphahani A., Cooper E. and Danesi P. R.: *J. Radioanal. Nucl. Chem.*, Articles, **162**, No.2 (1992) 307-323
3. BjØrnstad H. E., Lien H. N., Yu Yu-Fu and Salbu B.: *J. Radioanal. Nucl. Chem., Articles*, **156**, No. 1 (1992) 165-173
4. Horwitz E. P., Chiarizia R. and Dietz M. L.: *Solvent Extr. and Ion Exch.*, **10**, pp. 313-336, 1992
5. Horwitz E. P., Dietz M. L. and Fisher D. E.: *Solvent Extr. Ion Exch.*, **8** . Iss. 4 and 5, 1990
6. Huntley M.W.: U.S. Patent 5,494,647/A/. 27 Feb 1996.
7. Vincze Á., Nagy L Gy., Solymosi J., Hartmann P. and Erd• s E.: *Hungarian Chemical Series*, **100**. 1994. 4. sz (in hungarian).
8. Vincze Á., Hartmann P., Volent G., Solymosi J. and Nagy L.G.: A quick method for the determination of $^{90}$Sr in aqueosus samples. *Austrian - Italian - Huangarian Radiation Protection Symposium*: Obergurgl Austria, 28-30, April 1993. *Proceedings*, **2**. P. 267.
9. Sato T.: *Radioanal. Nucl. Chem., Articles*, **158**, No. 2. (1992) 391-399.
10. Bem H., Bakir Y. and Shuker S. M.: *J. Radioanal. Nucl. Chem.*, **147**. (1991), 263.
11. Kovalancik J. and Galova M.: *J. Radioanal. Nucl. Chem.*, **162**. (1992), 47.
12. Kessler M. J.: *Scientific News Letter*, **1**(2), (1990)
13. Dietz M. L., Horwitz E. P., Nelson D. M. and Wahlgren M. A.: *Health Phys.*, **61**. No. 6, pp. 871-877, 1991.
14. Chang T.M., Chen S.C., King J.Y. and Wang S.J.: *J. Radioanal. Nucl. Chem., Articles*, **204**, No. 2 (1996) 339-347.

# METHODS OF RETROSPECTIVE ESTIMATION OF THE ABSORBED DOSE IN ENVIRONMENTAL OBJECTS AND THE HUMAN BODY

# EVALUATION OF POSSIBLE RADIATION EXPOSURES NEAR THE SEMIPALATINSK NUCLEAR TEST SITE

S. L. SIMON
*Board on Radiation Effects Research,*[a]
*National Research Council,*
*Washington, DC USA*

## Abstract

To assess the external exposures that were likely received in villages near the Semipalantinsk nuclear test site in Kazakhstan, a dosimetry method consistent with available data is necessary. This paper presents calculations to estimate the integral external exposure by using two approaches: 1) historical exposure rate data, and 2) contemporary measurements of $^{137}Cs$ inventory in the soil. The input data and principal assumptions are discussed as well as estimates of integral exposure received in the village of Dolon as a result of a test conducted in August 1949. The estimated exposures measured by both methods were substantially less than those reported in the literature.

## 1. Introduction

Over one hundred atmospheric tests were conducted by the Soviet Union between 1949 and 1962 in eastern Kazakhstan at the Semipalatinsk Test Site (STS) (Gusev *et al.* 1997). The approximate explosive yield of these tests was 6.4 MT, an amount 6 times greater than the explosive yield of the above ground tests at the Nevada Test Site and about 6% of the yield of the tests conducted in the Marshall Islands.

The first nuclear test conducted at the Kazakhstan site was a plutonium bomb and was exploded on 29 August 1949. The device was reported to be a copy of the U.S. "Fat Man" design (Mikhailov 1996) and yielded about 22 kT of explosive power. Fallout dispersed from the test has been reported to have exposed residents of several villages located along a narrow trajectory to the northeast of the STS. Still today, fears concerning radiation effects continue to disturb many residents of villages near the site (Williams 1997).

The purposes of this report are two-fold. First, the essentials of two methodologies for calculating external exposure and the results from both calculations will be presented. Second, the calculation of externally delivered dose to inhabitants of Dolon, a village near the STS, will be compared with the dose values reported in the literature.

[a] Note new address in List of Authors.



*S.S. Hecker et al. (eds.), Nuclear Physical Methods in Radioecological Investigations of Nuclear Test Sites,* 153-159.
© 2000 *Kluwer Academic Publishers. Printed in the Netherlands.*

## 2. Methods

### 2.1. DOSIMETRY

The methods described in this report for retrospective dose assessment of the free-in-air exposure rate from nuclear weapons test fallout follows the methods developed by Hicks (1981; 1982) and Anspaugh and Church (1986). Variations on these methods were implemented in United States dose reconstruction studies (Beck and Anspaugh 1990; Simon *et al.* 1995) and later discussed by Beck (1996).

Though numerous tests at the STS may have resulted in deposition of caesium in Dolon, the dose calculated in this report assumes that all of the present-day caesium inventory (except for a global fallout contribution) was deposited following the test in August 1949. The calculations depend on two different schemes: 1) integrating historical exposure rate data collected soon after the test, and 2) converting contemporary measurements of soil caesium to the exposure rate following the test, and then integrating this measurement to infinity.

#### *2.1.1. Using Historical Exposure Rate Data*

Empirical exposure rate data are the most direct and easy to use type of data to determine integral exposures because the data need only to be integrated from time-of-fallout arrival to infinity and converted to a shielded whole-body dose. Often the exposure rate is measured later than when the deposition first occurs; thus, a decay correction is generally needed from the time of measurement to the time of deposition. The potential weakness of this approach results from the imprecision of the measurement and the difficulty in determining the applicability of the data (location of measurement relative to residences, etc.), both of which are extremely difficult to determine years after the events took place.

The basic equation for calculating external dose using this method is:

$$\text{Dose} = \left[\left\{\left(\frac{X}{\dot{X}_{12}}\right)\dot{X}_{12}\,\frac{D}{X}\,SF\frac{T_{in}}{24}\right\} + \left\{\left(\frac{X}{\dot{X}_{12}}\right)\dot{X}_{12}\,\frac{D}{X}\,\frac{(24-T_{in})}{24}\right\}\right] \tag{1}$$

where[1],

Dose (cSv) is the whole body effective dose,

$X/\dot{X}_{12}$ (units of R per R $h^{-1}$ at H+12) is the integral of the ten-component exponential decay equation from fallout time-of-arrival to infinity,

$\dot{X}_{12}$ (units of R $h^{-1}$) is the exposure rate at 12-h post-detonation,

D/X (units of cSv $R^{-1}$) is the factor converting exposure (Roentgen) to whole-body effective dose (cSv),

---

[1] The reader will note the use of a mixture of conventional and SI units owing to the situation in which contemporary measurements have been reported in SI units and factors drawn from the literature are shown in their original form, often in conventional units. The appropriate conversions have been made in the calculations.

SF is the fractional value of the outdoor exposure rate received indoors (building shielding factor), and

$T_{in}$ (units of hr) is the number of hours per day spent indoors.

*2.1.2. Using Contemporary Measurements of $^{137}Cs$*

Contemporary measurements of caesium inventory can also be used to estimate integral exposure. For this calculation, four additional steps are required:

1. Correct the caesium data for decay to August 1949;
2. Deduct the caesium contributed by global fallout;
3. Convert the net caesium to exposure rate at 12-h post-detonation;
4. Integrate the exposure rate to infinity and correct for body and building shielding. Possible weaknesses of this approach include the loss of caesium from the soil (due to disturbance, etc.). This loss would result in underestimating the exposure and in some additional necessary assumptions.

The basic equation for calculating external dose using this method is the same as that in Eq. (1) above, except that $\dot{X}_{12}$ is determined from contemporary soil measurements of $^{137}Cs$ rather than from the direct measurement of the exposure rate:

$$\dot{X}_{12} = \frac{AI_{net}}{ND} \tag{2}$$

where,

$AI_{net}$ (units of $\mu Ci\ m^{-2}$) is the aerial inventory of $^{137}Cs$ at time of deposition (determined by decay correcting measurement data to date of deposition and deducting global fallout component), and

ND (units of $\mu Ci\ m^{-2}$ of $^{137}Cs$ per $R\ hr^{-1}$ at H+12) is the normalised deposition factor.

## 2.2. INPUT DATA

The input data for both calculation methods are summarised in Table 1 and discussed briefly below. For the method using historical data on exposure rate, only a single measurement of exposure rate in the vicinity of Dolon is available: 0.1 $R\ h^{-1}$ at H+173 h.[2] The decay correction of this data to 12 h post-detonation (H+12) is an important step to achieving reliable exposure estimates. Most historical documents cite a variation of the Way-Wigner power law $t^{-1.2}$ for performing decay corrections of fallout, though the decay rate exponent is known to vary between about 1.2 and 1.4 (see Anspaugh and Church 1986), depending on the time after the explosion. In this work, a ten-term exponential equation was used. The fit to the multiterm exponential was provided by Henderson (1991) using empirical exposure rate data from Hicks (1985) for the United States Trinity test.

For the method using contemporary data on the soil caesium, measurements of $^{137}Cs$ inventory were provided by one component of the International Association for the Promotion of Cooperation with Scientists from the Newly Independent States of the Former Soviet Union (INTAS), a study funded by the European Union. The $^{137}Cs$ aerial

[2] K. I. Gordeev, 1998. Communication to NCI Workshop on Kazakhstan Dosimetry, Rockville, MD, 1998.

inventory data obtained in Dolon by the University of Salzburg (as part of INTAS) were derived from 18 vertical profiles sampled to 30-cm depths. Basic statistical analysis indicated the data were near normally distributed with a mean of 4790 Bq, a median of 5370 Bq, and a standard deviation of 2960 $Bq/m^2$. The maximum value observed was about 10 $kBq/m^2$; the minimum value was 0.15 $kBq/m^2$. The maximum sampled valued by the INTAS project agrees well with the data reported by other research groups.

Caesium-137 deposition from global fallout was derived from the United Nations Scientific Committee on the Effects of Atomic Radiation (UNSCEAR) report (1993) that used data originally published by the Health and Safety Laboratory (HASL) of the United States Energy Research and Development Administration (1977). The reported deposition of $^{90}Sr$ at 50°-60° N latitude was 2890 $Bq/m^2$. The fission yields of $^{137}Cs$ and $^{90}Sr$ were reported to be 5.57% and 3.5%, respectively. Thus, $^{137}Cs$ ($Bq/m^2$ in 1996) ≈ (5.57/3.5) x 2890 x exp[-(0.693/30) x 19] = 2965 $Bq/m^2$ ≈ 3000 Bq/m2. The estimated value of $^{137}Cs$ deposition from global fallout (after being decay corrected to 1996) was greater than 30% of the values measured by the INTAS project.

The time-dependent relationships between total fallout deposition, individual radionuclide deposition densities, and exposure rates (for all fission and activation productions) were developed by Hicks (1981; 1982) for United States tests. These relationships were based on classified bomb construction information, fission yields, and published exposure rates. Hicks proposed that the ratio of volatile to refractory elements (V/R) in the nuclear cloud would decrease from unity at the location of the detonation to a value of 0.5 at a distance of 260 km and beyond. The value of V/R that is assumed in the calculation will affect the exposure rate derived from the $^{137}Cs$ measurements.

The relationship between exposure rate and deposition (Bq $m^{-2}$ per mR $h^{-1}$ at H+12) is termed, for convenience, the normalised deposition factor because the deposition is normalised to the exposure rate at 12 hours after detonation or H+12. Typical values derived by Hicks (1981) for $^{137}Cs$ are of the order of 1 x $10^{-3}$ μCi $m^{-2}$ per mR $h^{-1}$ at H+12 for V/R = 0.5 and 9 x $10^{-4}$ μCi $m^{-2}$ per mR $h^{-1}$ at H+12 for V/R = 1. Because of Dolon's close location to the STS, an appropriate value for the normalised deposition factor is presently believed to be significantly less than the values used for dose reconstruction in most United States studies.

Fallout-transit time determines the degree of decay of the extremely short-lived activation productions in fallout. Reliable estimates of fallout-transit time, thus, partially determine the precision of estimated integral exposures. Rosenson et al. (1996) and Gusev et al. (1997) reported the wind velocity at the time of the 1949 test (29 August 7:00 AM) to be from 45 to 50 km/h and that a radioactive cloud arrived in Dolon and other nearby villages 2 hours after the explosion. In this work, we used a value of 2.5 h, recommended by Gordeev (personal communication).

Shielding factors for buildings account for the reduction of the free-in-air exposure due to the effectiveness of building material for reducing the exposure rate indoors and also depend on the construction materials, size, etc. of the buildings. In these calculations, the indoor exposure rate was assumed to be one-third of the outdoor value.

Factors for conversion of exposure (R) to equivalent dose (Sv) were taken from the International Commission on Radiological Protection (ICRP) (1987). The value used (see Table 3a of ICRP) was 0.72 cSv/R, which is specific for uniform rotational irradiation.

## 3. Findings and Discussion

### 3.1. INTEGRAL EXPOSURES

Under ideal circumstances, integral exposures calculated directly from exposure rate data and as derived from measurements of soil-radioactivity inventories should agree. However, in the case reported in this paper, poor agreement was noted between the two methods. This discrepancy was particularly apparent when the nominal value of the normalised deposition factor was used (see Table 1). Further discussion on this point follows.

TABLE 1. Summary of parameter values used in dosimetric calculations

| Parameter | Method | |
|---|---|---|
| | Based on exposure rate | Based on soil $^{137}Cs$ inventory |
| Basic input data | 0.1 R $h^{-1}$ at H+173 h | 10 000 Bq $m^{-2}$ (max)<br>5 370 Bq $m^{-2}$ (median) |
| Global fallout deposition | (not needed) | 3 000 Bq $m^{-2}$ |
| Fallout time-of-transit | 2. 5 h | 2.5 h |
| Integration of 10-component exponential, from H +2.5 hr to ∞ | 79.7 R per R $h^{-1}$ at H+12 | 79.7 R per R $h^{-1}$ at H+12 |
| Normalised deposition factor for $^{137}Cs$ | (not needed) | 1 x $10^{-3}$ μCi $m^{-2}$ per mR $h^{-1}$ at H+12 (nominal value)[a] |
| Time spent indoors/outdoors | 8 h/16 h | 8 h/16 h |
| Indoor shielding reduction | 1/3 (unitless) | 1/3 (unitless) |
| cSv/R | 0.72 | 0.72 |

[a]Nominal value was adjusted in calculations (see Discussion in text of paper).

#### *3.1.1. Integral Exposure Calculated from Historical Exposure Rate Data*

The estimate of exposure rate at 12 hours after detonation was calculated to be 1.6 R/h when using the multi-component exponential. By using that value as input, the integral external dose received by Dolon inhabitants was estimated to be 93.5 cSv (rad) with no building shielding and 72.1 cSv assuming 8 hours spent indoors daily.

#### *3.1.2. Integral-Exposure Estimate Calculated from Soil Sampling*

Using contemporary measurements of soil $^{137}Cs$ to estimate the external exposure received by Dolon residents generally resulted in overall lower estimated doses, in part, because some of the soil caesium values were very low. Using the maximum observed caesium value and the nominal value of the normalised deposition factor (1 x $10^{-3}$ μCi/ $m^{-2}$ per mR $h^{-1}$ at H+12) resulted in an estimate of 32.2 cSv with no building shielding, and in an estimate of 24.8 cSv assuming 8 hours spent indoors daily. The median observed soil inventory (5370 Bq $m^{-2}$) yielded integral-exposure estimates of 10.9 and 8.4 cSv, with and without building shielding.

Because of Dolon's close proximity (~120 km) to the STS, the ratio of volatile to refractory radionuclides could be much less than 1.0 and likely less than 0.5. The net effect would be a larger exposure rate per unit of deposited $^{137}Cs$ than would be expected at distances farther from the test site. A value of the normalised deposition factor equal to 3.43 x $10^{-4}$ (about 35% of the nominal value) adjusts the integrated exposure to equal the value calculated with the historical exposure rate measurement: 93.5 cSv with no building shielding and 72.1 cSv assuming 8 hours spent indoors daily. Using the reduced value of the normalised deposition factor, the median soil caesium inventory value yielded integral exposures of 31.7 cSv and 24.4 cSv, with and without building shielding.

## 3.2. VARIATION AND UNCERTAINTY OF EXPOSURES AMONG RESIDENTS

The variation of integral exposures likely received in Dolon could best be estimated using a range of reasonable input data. However, because of the paucity of exposure rate data, it is presently not possible to determine the variation of doses by using the exposure rate method. When the soil inventory is used as a basis for the calculations, it shows that about 30% of the population might not have been exposed to the 1949 test as the inventory at 30% of the locations is not greater than that of the global fallout deposition. Consequently, the range of estimated exposures from this calculation extends from zero to the value calculated using the maximum soil concentration.

It is not known with any certainty the reasons why the soil profiles exhibited the range of observed inventories. Two possible reasons are that the data actually reflect the variation of deposition within Dolon, or that some of the profiles may have suffered losses of radioactivity during the intervening years as a result of human and animal disturbance and/or the natural weathering process.

The estimation of uncertainty of exposures is closely coupled to the question of variability as well as to the lack of knowledge about the critical parameters in the calculation. Whereas the uncertainty of the decay rate is usually of greatest concern, the multi-component exponential fitted to data from the Trinity test is believed to be reasonably applicable to the Dolon test. Hence, the greatest uncertainties are likely due to the lack of information about applicability and precision of the single exposure rate measurement, and of the appropriate value of the normalised deposition factor.

## 3.3. COMPARISON WITH LITERATURE ESTIMATES

The calculations presented here for the likely values of the integral exposure received by Dolon residents do not agree well with those published previously. In particular, the two methods used here indicate that integral exposures (including assumptions on building shielding) were likely around 70 cSv. Smaller exposures may have been received by some residents, depending on the degree of protection by buildings and on the variations in deposition within the village area. There is no evidence from either the exposure rate or the soil inventory methods that external exposures were likely to have been larger than the estimate of 70 cSv. Exposures to Dolon inhabitants from the 1949 test were reported by Rosenson *et al.* (1996) and Gusev *et al.* (1997) to be 2 Sv (200 rem) without shielding and 1.6 Sv with shielding. The exposure estimates presented here are, in general, less than half of those values. It should be understood, however, that the uncertainty in the values reported here and those published earlier have not been fully evaluated.

## 4. Acknowledgements

Appreciation is extended to Frederik Steinhausler of the University of Salzburg for making the soil caesium data available for this analysis. Appreciation is extended to NATO for conference and travel support, to Los Alamos National Laboratory for organisational support, and to the Kazakhstan Nuclear Centre for graciously hosting the workshop.

**Author's Note:** This manuscript does not represent the opinion of the National Academy of Sciences or the National Research Council.

## 5. References

1. Anspaugh, L. R.; Church, B. W. Historical estimates of external $\gamma$ exposure and collective external $\gamma$ exposure from testing at the Nevada Test Site. I. Test series through Hardtack II, 19568. Health Physics 51:35-51. 1986.
2. Beck, H. L. Reconstructing source terms from environmental data. Environmental Dose Reconstruction and Risk Implications. Till, J. E. ed. Bethesda: National Council on Radiation Protection and Measurements. 17: 79-91. 1996.
3. Beck, H. L.; Anspaugh, L. R. Development of the County Database: Estimates of Exposure Rates and Times of Arrival of Fallout in the ORERP Phase II Area. Comparison With Cumulative Deposition Estimates Based on Retrospective and Historical Soil Samples. U.S. Department of Energy, Nevada Operations Office, Las Vegas, NV 89114. NVO-320. 1990.
4. Gusev, B. J.; Abylkassimova, Z. N.; Apsalikov, K. N. The Semipalatinsk nuclear test site: a first assessment of the radiological situation and the test-related radiation doses in the surrounding territories. Radiat Environ Biophys 36:201-204. 1997.
5. HASL. Final Tabulation of monthly Sr-90 fallout: 1954-1976. NY: Health and Safety Laboratory, U.S. Energy Research and Development Administration. 1977.
6. Henderson, R. Approximation of the Decay of Fission and Activation Product Mixtures. Los Alamos National Laboratory. LA-11968-MS. 1991.
7. Hicks, H. G. Results of Calculation of External Gamma Radiation Exposure Rates from Fallout and the Related Radionuclide Compositions, Parts 1-8. Lawrence Livermore Laboratory, Livermore, CA. UCRL-53152. 1981.
8. Hicks, H. G. Calculation of the Concentration of any Radionuclide Deposited on the Ground by Off-Site Fallout from a Nuclear Detonation. Health Physics 42:585-600. 1982.
9. Hicks, H. G. Results of Calculations of External Gamma Radiation Exposure Rates from Fallout and the Related Radionuclide Compositions - the Trinity Event. Livermore, CA: Lawrence Livermore National Laboratory. UCRL-53705. 1985.
10. ICRP. Data for Use in Protection Against External Radiation. ICRP Publication 51 Oxford: Pergamon Press. 1987.
11. Mikhailov, V. N., ed. USSR Nuclear Weapons Tests and Peaceful Nuclear Explosions: 1949 through 1990. The Ministry of the Russian Federation for Atomic Energy and Ministry of Defense of the Russian Federation. 1996.
12. Rosenson, R. I.; Tchaijunusova, N. J.; Gusev, B. I.; Katoh, O.; Kumura, A.; Hoshi, M.; Kanada, N.; Satow, Y. Late effects of exposure to ionizing radiation - Studies of the residents in the Semipalatinsk area. Proc. Hiroshima Univ. RINMB 36:177-253. 1996.
13. Simon, S. L.; Till, J. E.; Lloyd, R. D.; Kerber, R. L.; Thomas, D. C.; Preston-Martin, S.; Lyon, J. L.; Stevens, W. The Utah Leukemia Case-Control Study: Dosimetry Methodology and Results. Health Physics 68(4):460-471. 1995.
14. United Nations Scientific Committee on the Effecs of Atomic Radiation. Sources and Effects of Ionizing Radiation, UNSCEAR report to the General Assembly with scientific annexes. New York: United Nations. 1993.
15. Williams, D. Legacy of Soviet nuclear tests haunts Khazakhstan. Washington, D.C.: The Washington Post. November, 1997.

# TECHNICAL BASIS OF DOSE RECONSTRUCTION

L.R. ANSPAUGH
*Radiobiology Division*
*University of Utah*
*729 Arapeen Drive*
*Salt Lake City, UT 84108-1218*

## Abstract

There is a long-standing interest in the reconstruction of radiation doses received by populations from the testing of nuclear weapons or from the release of radionuclides associated with the production of special nuclear materials. These ongoing methods of dose reconstruction have evolved over time and have been quite successful in supporting epidemiologic studies and in examining issues related to social justice. Methods of dose reconstruction vary from the analysis of a sample from the exposed individual to the analysis of a sample from the individual's home, the analysis of environmental residues (such as long-lived radionuclides), or to simpler calculations that are highly dependent upon models. The more reliable methods for reconstructing doses from nuclear weapons tests have focused on using historical measurements of external gamma-exposure rate and on historical or contemporary measurements of the deposition density of long-lived radionuclides. This paper describes in detail the general method of dose reconstruction used in the U.S.A. for doses to residents near the Nevada Test Site.

## 1. Introduction

Over approximately the last 50 years, the reconstruction of radiation doses to hypothetical or actual persons has become a major scientific activity.[1] Many dose-reconstruction activities have been undertaken to support companion epidemiologic studies of radiation risk or to help achieve social justice [1].

The first major dose-reconstruction study was undertaken for the survivors of the atomic bombings in Japan. Three major steps have occurred in this still-continuing activity: the establishment of tentative 1957 (T57) doses [3, 4], tentative 1965 (T65) doses [5, 6], and the more realistic doses established by the Dosimetry System 1986 [7].

---

[1] Much of the introductory material presented here has been modified from References [1] and [2].



*S.S. Hecker et al. (eds.), Nuclear Physical Methods in Radioecological Investigations of Nuclear Test Sites*, 161-172.
© 2000 *Kluwer Academic Publishers. Printed in the Netherlands.*

This long-running activity is still not finished, as newer and ever-more sophisticated measurements of residual activities indicate major discrepancies in the neutron doses for the survivors in Hiroshima [8].

In the United States many dose-reconstruction activities have been undertaken to assess the dose and possible risks to the general population from the continental testing of nuclear weapons and from the operation of the many sites involved in producing special nuclear materials. A brief indication of these studies is given in Table 1.

TABLE 1. Major dose-reconstruction studies conducted in the U.S.A. and concerned with the general population living in the vicinity of sites for the production of special nuclear materials or the testing of nuclear weapons

| Site | Object of assessment | References |
|---|---|---|
| **Completed** | | |
| Nevada Test Site | Hypothetical individuals | [9-11] |
| | Collective | [12, 13] |
| | Specific individuals | [14, 15] |
| Hanford Works | Hypothetical individuals | [16-18] |
| **In process** | | |
| Hanford Works | Specific individuals | [19] |
| Fernald Feed Materials Production Center | | [20] |
| Rocky Flats Plant | | [20] |
| Oak Ridge Site | | [20] |
| Idaho National Engineering Laboratory | | [20] |
| Savannah River Site | | [20] |

In Europe and Asia the largest activity has been associated with reconstructing and projecting doses from the Chernobyl accident that occurred in 1986 [21-26].

Finally, with the breakup of the Soviet Union, it has become known that large releases and exposures had occurred from the operation of the Mayak Production Facility in the Urals region of Russia. In this area, the dose-reconstruction study for the Techa River residents had been ongoing since the early 1950s; current results have been published [27-29], and a project to improve the dosimetry database is ongoing [30].

## 2. Methods of Dose Reconstruction

Over the years the many methods of dose reconstruction devised and implemented range widely from old to new, from inexpensive to expensive, from calculational to measurement based, from specific to general, and from roughly approximate to highly accurate. In general, highly accurate methods tend to be rather specific and expensive.

Dose-reconstruction methods have been discussed and summarised recently elsewhere [1, 31, 32]; a brief description of the major methods follows.

An outline of the major methods of dose reconstruction is indicated in Table 2. This table is modified from Reference [1] to present more methods, particularly those more appropriate for workers, as opposed to including only those appropriate for members of the general population. The primary difference among the two population groups is that workers generally were in more controlled environments than were members of the general population and the workers had more extensive monitoring of both external and internal exposures.

TABLE 2. Methods of dose reconstruction. The methods are arranged roughly in an hierarchy moving downward from reliable and preferred to approximate

| |
|---|
| Individual "biologic" analysis |
| Bioassay (measurement of radionuclides in urine and/or feces) |
| Whole- or partial-body counting to determine radionuclide content |
| Measurement of samples collected at autopsy |
| Analysis of stable, reciprocal chromosome translocations in circulating lymphocytes |
| Electron paramagnetic analysis of teeth |
| Individual film-badge dosimeters |
| Individual "environmental" analysis |
| Area film badges |
| Area air monitors |
| Thermoluminescent analysis of samples from home environment |
| Analysis of environmental residues |
| Radionuclide-deposition densities, past or present |
| External gamma-exposure rates, past |
| Activation products (for neutrons), e.g., $^{36}Cl$ and $^{63}Ni$ |
| Known releases plus models |
| Models of radionuclide transport in air or water |
| Models of radiation transport in air or water |
| Inferred releases plus models |
| Models of radionuclide transport in air or water |
| Models of radiation transport in air or water |
| Known or inferred releases, coupled to global dose factors |

Of course, the most preferred method of dose reconstruction would be "biologic" and would allow a direct indication from the individual of what exposure had occurred. For radionuclides such as $^{90}Sr$ and $^{239}Pu$, which stay in the body for a very long time, it is possible to ascertain the systemic burden based upon bioassay results and, for $^{90}Sr$, on whole body counting [33] of the bremsstrahlung from the beta emitted by $^{90}Y$. Samples, particularly of bone, collected at autopsy can also be measured for the content of long-lived radionuclides, again more usefully for $^{90}Sr$ and $^{239}Pu$. Even more valuable would be the analysis of a sample extracted from the person. This sample could be analysed to give a more direct measure of the dose actually received. There is not yet a perfect dosimeter of this type, but two relatively recent developments show great promise.

The first method is the analysis of stable, reciprocal translocations in circulating lymphocytes. Such translocations supplement the well-known measurement in circulating lymphocytes of chromosome dicentrics, an analysis that is very useful soon after a significant exposure has occurred. However, the analysis of dicentrics is not useful over long time periods following exposure. On the other hand, it has been demonstrated that reciprocal translocations in chromosomes of the same circulating lymphocytes are stable over at least several decades [31]. The most convenient method of assay is "chromosome painting" with fluorescence *in situ* hybridisation (FISH) [34], though this method is time consuming and expensive. The response of the assay to tritium betas and to gamma rays of various energies has been quantified, but its response to many incorporated radionuclides has not been quantified. The sensitivity of the FISH assay is on the order of 100 mGy, depending upon the number of cells that are counted and upon the individual's age, factors that determine the background incidence of translocations.

The second "modern" method of individual-dose determination is the electron paramagnetic resonance (EPR) analysis of dentine [31, 35]. This method has developed rapidly as a result of intense interest in its use following the accident at Chernobyl. However, there are known problems with the method, such as the interference of the signal by ultraviolet light for the front teeth and individual differences in the sensitivity of response, though methods are rapidly being found to solve these problems. The application of the technique requires the extraction of a tooth, and the analysis is expensive and time consuming, if the problems mentioned above are addressed adequately. The sensitivity of the method is around 50 mGy. The method would not work well for individuals who have had extensive exposure to dental X-rays, although such soft X-ray exposure can be quantified by careful work with depth profiles of dose within an individual tooth.

Most workers in the nuclear industry, even from the earliest days, wore individual dosimeters of some type designed to record external gamma and beta dose. Typically, these individual dosimeters were film badges during the early periods of higher exposures; the design of such film badges has been improved over the years to allow for the differential energy response of film. More recently, thermoluminescent dosimeters have been used. Such individual badges provide extremely valuable and reliable information, but in general the resulting data cannot be used directly without careful consideration of bias and uncertainty in the resulting estimates of dose, especially when dose for individual organs is needed [36].

If direct analysis of the individual is not possible, the next more reliable method would be analysis of some material taken from within the individual's environment. For a worker this might be using area-radiation monitors, such as film badges, or using area monitors that record the concentration of radionuclides in air. For individuals in the general population, one technique that has been used quite successfully is the thermoluminescent analysis of materials from homes. To be useful the thermoluminescent signal must have been "zeroed out" at some recent time before exposure; this is usually done by the firing of bricks and porcelain during the manufacturing process. Such analyses have proved useful for dose reconstructions for the survivors of the atomic bombings in Japan [37], for the population living nearby the Nevada Test Site [38], and for the population living downstream of the Mayak plutonium-production facility in the southern Urals [39, 40]. The sensitivity of the method is approximately 50 mGy.

The analysis of environmental residues is a well-established technique that has been used since the early days of the study of global fallout and of the resulting dose determination to man [41]. A major component of such studies was to measure the deposition rate of $^{90}Sr$ through a variety of collectors and to measure the cumulative deposition density of $^{90}Sr$ by analysing soil samples. The deposition density of all other radionuclides, even though they might have decayed long ago, could then be inferred on the basis of what was known about the source term of the releases. Modern calculations using gamma-ray transport codes have allowed the conversion from radionuclide-deposition density to external gamma-exposure rate [42] or external dose rate [43]. Thus, provided that the relative source term of any release is known, it is possible to infer the deposition densities of all radionuclides and the resulting external gamma-exposure rate as a function of time, if the deposition density of any radionuclide is known or if the external gamma-exposure rate was measured at some early time. This technique was used extensively in the dose reconstruction for the residents living near the Nevada Test Site [9, 10, 44, 45]. The accelerator-mass-spectrometry analysis of $^{129}I$, half-life of 16 million years, as a surrogate for $^{131}I$, half-life of 8 days, has also been demonstrated as a useful tool for the reconstruction of doses to the thyroid as a result of the Chernobyl accident [46].

The analysis of the neutron-activation product, $^{36}Cl$, in samples of concrete taken from Hiroshima has recently provided information on the flux of thermal neutrons at Hiroshima [8]. The next step in resolving the current discrepancy (measurements versus calculations) in neutron doses at Hiroshima is to analyse some product that would respond only to high-energy neutrons. At the present time, $^{63}Ni$ is the leading candidate for this analysis, and an analytical system has been demonstrated [31, 47].

The remaining methods listed in Table 2 are not usually reliable in episodic releases [1]. However, when releases occur over long time periods, the ability of atmospheric transport models to provide approximately accurate time-integrated predictions is much improved. This method was used for the calculations of thyroid dose from $^{131}I$ from the Hanford Works; such calculations depended entirely upon the laborious definition of the source term on the basis of process records and the use of an atmospheric transport model [16]. In this case, there was no other viable choice, as no known environmental residues were capable of being a surrogate for $^{131}I$. (For this application $^{129}I$ is not a viable surrogate for $^{131}I$, as the filters installed at later times to trap $^{131}I$ eventually released the entrained $^{129}I$.) A river model was also used to calculate the dose from radionuclides released to the water at the Hanford Works [17], as the measurements available were not adequate alone to support an assessment. Sometimes it may be adequate to calculate collective dose only based upon an inferred source term and the use of "global dose factors." Such techniques are discussed in References [48] and [49]. If the latter techniques are used, it is not possible to calculate individual doses.

## 3. Lessons Learned from Past Dose Reconstructions

Over the years the many dose-reconstruction studies have become increasingly sophisticated as newer methods of reconstruction, verification, and validation have been developed. Many of the "tools" in current use are of fairly recent development, and this development process will continue.

In general, there is no "right" method of dose reconstruction. Methods tend to be site- or problem-specific and tailored to the data available to support the reconstruction. In fact, one might say that the first rule of dose reconstruction is to make maximum use of the data available. Real measurements are extremely useful, particularly in supporting the credibility of the reconstruction. If the measurements themselves are inadequate to support the reconstruction, then the measurements must be saved for validation purposes. Validation is approached by comparing predicted values (of air concentration, deposition density, concentration in vegetation, etc.) to measured values, by comparing several independent types of measurements, and by checking for biological plausibility. (That is, are the predicted doses consistent with the medical observations or lack of observations?)

Methods that have been used vary from completely theoretical (or calculational) to those based on measurements only. It is unlikely that a wholly theoretical study will yield results of sufficient accuracy, as a large number of assumptions are typically required, and there are too many variables with uncertain influence.

The more successful applications use a combination of approaches, devote a major fraction of effort to verification and validation, and provide estimates of the central value of reconstructed doses and uncertainty. Dose reconstruction is a very difficult task not generally accomplished through the application of textbook methods. An important point is that dose reconstruction should be a continuing activity over the life of any associated epidemiologic study. It is important to note that there is a deplorable tendency to view dose reconstruction as something trivial and easily accomplished within a fraction of the time afforded to an epidemiologic study.

## 4. Reconstruction of Doses from Tests of Nuclear Weapons-Related Devices

As the main topic of this Workshop is nuclear test sites, it is interesting to consider in more detail the reconstruction of doses to persons living downwind of sites where nuclear weapons have been tested. As mentioned above, this has been a major research activity in the U.S.A. [9-15, 38, 42, 44, 45]; similar studies have also been conducted in the former Soviet Union [50-52], although results are still incomplete for the residents in Kazakhstan. In general, the better bases of dose reconstruction are historical measurements of external gamma-exposure rate with hand-held or airborne instruments and historical or contemporary measurements of the deposition density of long-lived radionuclides. Attempts to use atmospheric transport models alone [53] to delineate deposition patterns are not successful, as is the case with episodic events, in general. Measurements of the thermoluminescence of quartz samples from bricks, etc. and measurements of stable chromosome translocations in theory would be very useful.

However, these are typically only available in quantities sufficient to be helpful for validation studies [38, 54].

Given that the useful starting point for dose reconstruction is an historical measurement of external gamma-exposure rate or of the deposition of one or more radionuclides, an essential tool is the ability to convert between external gamma-exposure rate and deposition density for all significant radionuclides. For the U.S. studies, Beck [42] used a gamma-ray-transport code to provide exposure-rate-conversion factors for all fission- and activation-product radionuclides of interest. These data, combined with radiochemical data [55] collected for all U.S. atmospheric tests in Nevada were used to develop shot-specific, time-dependent catalogues [44, 56] of total exposure rate and radionuclide-deposition density. All data were normalised to an exposure rate of 1.0 mR per hour, twelve hours after the time of detonation.

The decay of the external gamma-exposure rate does not follow accurately the frequently assumed power function of elapsed time [44, 56, 57]. Rather, a better approximation is achieved by using a sum of ten exponential functions, which were specifically fit [58] to the more complete data for each test [44, 56]. Thus, external exposure, $X$, is calculated [9] as

$$X = \int_{a}^{\infty} \dot{X}(12) \cdot S(t) \cdot \sum_{i=1}^{10} c_i e^{-\lambda_i t} \cdot dt \;, \qquad (Eq.1)$$

where $\dot{X}(12)$ is the external gamma-exposure rate at twelve hours after the event at a particular location,

$S(t)$ is the time-dependent shielding factor,

$a$ is the time of arrival of fallout at the location of interest, and

$c_i$ and $\lambda_i$ are fit coefficients that describe the decay of the external gamma-exposure-rate field.

The function $S(t)$ can be generalised to a single value, or, more accurately, can be varied every few hours to allow for the detailed lifestyle of a hypothetical or real person. In any case, it is necessary to know the shielding characteristics of homes typical for a given location.

The calculation of internal dose from the ingestion of contaminated foodstuffs is much more complicated, as the radioecological movement of radionuclides must be considered as well as the age-dependent food-consumption habits of the population of interest. A general equation [10] for a modular approach developed in the U.S. for the calculation of dose, $D$, from ingestion is

$$D = \dot{X}(12) \cdot P \cdot I \cdot F_g \;, \qquad (Eq.2)$$

where $P$ is the normalised deposition of the radionuclide with the normalisation to unit-exposure rate at 12 hours after detonation (μCi per $m^2$ per mR per hour at 12 hours),

$I$ is the integrated intake by ingestion of the radionuclide per unit deposition (μCi per μCi per $m^2$), and

$F_g$ is the ingestion dose coefficient for the radionuclide of interest (rad per μCi).

The values of $P$ for U.S. shots are taken from the tabulations prepared by Hicks [44, 56]. The values of $F_g$ were calculated for 22 organs and four age groups specifically for this application [59]; more recent results of the International Commission on Radiological Protection [60] may also be used.

The most difficult aspect of the calculation is the evaluation of the integrated intake, $I$. For U.S. calculations, a time-dependent model [61-64] has been developed to treat the radioecological processes involved in the retention of radionuclides by vegetation, the subsequent movement of radionuclides through food chains, and the consumption rates by four different age groups. A very critical factor, and one to which the model is very sensitive, is the amount of the total deposition retained by vegetation. In U.S. terminology, this parameter is termed the mass-interception factor and is the amount of the total deposit retained by vegetation divided by the dry biomass. The value chosen for this function for U.S. calculations near the Nevada Test Site was typically 0.39 $m^2$ per kg [61]. For further locations, a distance- or time-dependent value of this parameter was used by Simon *et al.* [65] with a maximum value of 3.0 $m^2$ per kg.

In general, the dose from the ingestion of radionuclides with food is less than the dose from external exposure, with the exception of the dose to the thyroid from the ingestion of radioiodines [10].

Dose from the inhalation of radionuclides during the passage of the cloud (or from the resuspension of the deposited material) is typically small compared to the dose from the ingestion of radionuclides or from external exposure. For two events considered to represent extreme cases, the percentage of internal doses from inhalation compared to ingestion varied from a few percent to a few tens of percent [10].

Recently, the U.S. National Cancer Institute has sponsored a series of workshops that provided an opportunity to compare U.S. and Russian methods [50, 51] of calculating doses from nuclear weapons tests. Agreement was rather good on calculations of external dose but varied from reasonable to poor on calculations of internal dose (especially thyroid dose from $^{131}I$). The key difference in the methodologies was determined to be the value associated with the retention of radionuclides on vegetation and with its subsequent biological availability. For the first USSR test on 29 August 1949, the deposition at Dolon near the Semipalatinsk Polygon was high, with resulting external exposures of approximately 100 R. This high level of deposition was beyond the envelope of U.S. experience, and the nominal U.S. value for the mass-interception factor of 0.39 $m^2$ per kg appears to be much too high. There are future plans to combine the U.S. and Russian experiences to prepare a unified dose-assessment methodology.

## 5. Acknowledgements

The U.S. Department of Energy and the U.S. National Cancer Institute funded portions of work underlying this article. The author appreciates many useful conversations with Konstantin Gordeev, Harold Beck, André Bouville, and Steven Simon.

## 6. References

1. Anspaugh, L.R. (1996) Technical basis for dose reconstruction, in *Environmental Dose Reconstruction and Risk Implications,* National Council on Radiation Protection and Measurements, Bethesda, pp. 25–48.
2. Anspaugh, L.R. (1999) Overview on existing approaches to dose reconstruction—Recommendations for the Ukrainian–American Chernobyl Ocular Study (UACOS) in *Ocular Radiation Risk Assessment in Populations Exposed to Environmental Radiation Contamination,* Kluwer Academic Publishers, Boston, pp. 89–98.
3. Arakawa, E.T. (1960) Radiation dosimetry in Hiroshima and Nagasaki atomic-bomb survivors, *New Engl. J. Med.* **263**, 488–493.
4. Auxier, J.A. (1977) ICHIBAN: Radiation Dosimetry for the Survivors of the Bombings of Hiroshima and Nagasaki, U.S. Department of Energy, Washington.
5. Milton, R.C. and Shohoji, T. (1969) Tentative 1965 Radiation Dose Estimation for Atomic Bomb Survivors, Radiation Effects Research Foundation, Hiroshima, Technical Report 1–68, 1968.
6. Auxier, J.A. (1982) Development of the dosimetric program, T65D values, in *Reevaluations of Dosimetric Factors: Hiroshima and Nagasaki,* U.S. Department of Energy, Washington, CONF–810928, pp. 6–24.
7. Roesch, W.C. (ed.) (1987) US–Japan Joint Reassessment of Atomic Bomb Radiation Dosimetry in Hiroshima and Nagasaki, Radiation Effects Research Foundation, Hiroshima.
8. Straume, T., Egbert, S.D., Woolson, W.A., Finkel, R.C., Kubik, P.W., Gove, H.E., Sharma, P. and Hoshi, M. (1992) Neutron discrepancies in the DS86 Hiroshima Dosimetry System, *Health Phys.* **63**, 421–426.
9. Henderson, R.W. and Smale, R.F. (1990) External exposure estimates for individuals near the Nevada Test Site, *Health Phys.* **59**, 715–721.
10. Ng, Y.C., Anspaugh, L.R. and Cederwall, R.T. (1990) ORERP internal dose estimates for individuals, *Health Phys.* **59**, 693–713.
11. National Cancer Institute (1997) Estimated Exposures and Thyroid Doses Received by the American People from Iodine-131 in Fallout Following Nevada Atmospheric Nuclear Bomb Tests, U.S. Department of Health and Human Services, Washington.
12. Henderson, R.W. and Smale, R.F. (1992) Summary of Collective Dose from External Exposure, Los Alamos National Laboratory, Los Alamos, Report submitted to the Co-ordination and Information Centre, Las Vegas.
13. Whicker, F.W., Kirchner, T.B., Anspaugh, L.R. and Ng, Y.C. (1996) Ingestion of Nevada Test Site fallout: Internal dose estimates, *Health Phys.* **71**, 477–486.
14. Simon, S.L., Till, J.E., Lloyd, R.D., Kerber, R.L., Thomas, D.C., Preston-Martin, S., Lyon, J.L. and Stevens, W. (1995) The Utah leukemia case-control study: Dosimetry methodology and results, *Health Phys.* **68**, 460–471.
15. Till, J.E., Simon, S.L., Kerber, R., Lloyd, R.D., Stevens, W., Thomas, D.C., Lyon, J.L. and Preston-Martin, S. (1995) The Utah thyroid-cohort study: Analysis of the dosimetry results, *Health Phys.* **68**, 472–483.
16. Farris, W.T., Napier, B.A., Ikenberry, T.A., Simpson, J.C. and Shipler, D.A. (1994) Atmospheric Pathway Dosimetry Report, 1944–1992, Pacific Northwest Laboratories, Richland, WA, PNWD-2228 HEDR.
17. Farris, W.T., Napier, B.A., Simpson, J.C., Snyder, S.F. and Shipler, D.A. (1994) Columbia River Pathway Dosimetry Report, 1944–1992, Pacific Northwest Laboratories, Richland, WA, PNWD-2227 HEDR.
18. Farris, W.T., Napier, B.A., Ikenberry, T.A. and Shipler, D.B. (1996) Radiation doses from Hanford Site releases to the atmosphere and the Columbia River, *Health Phys.* **71,** 588–601.

19. Davis, S., Kopecky, K.J., Hamilton, T.E. and Amundson, B. (1999) Summary Final Report of the Hanford Thyroid Disease Study, Fred Hutchinson Cancer Research Center, Seattle, WA.
20. Miller, C.W., Smith, J.M. and Denham, L.S. (1994) Dose reconstruction studies at selected nuclear weapons facilities in the USA, in *Assessing the Radiological Impact of Past Nuclear Activities and Events,* International Atomic Energy Agency, Vienna, IAEA-TECDOC-755, pp. 79–85.
21. Balonov, M.I. (1996) Chernobyl dose for population of areas radiocontaminated after the Chernobyl accident, in *Environmental Dose Reconstruction and Risk Implications,* National Council on Radiation Protection and Measurements, Bethesda, pp. 207–243.
22. Likhtariov, I., Kovgan, L., Novak, D., Vavilov, S., Jacob, P. and Paretzke, H.G. (1996) Effective doses due to external irradiation from the Chernobyl accident for different population groups of Ukraine, *Health Phys.* **70**, 87–98.
23. Likhtarev, I.A., Kovgan, L.N., Vavilov, S.E., Gluvchinsky, R.R., Perevoznikov, O.N., Litvinets, L.N., Anspaugh, L.R., Kercher, J.R. and Bouville, A. (1996) Internal exposure from the ingestion of foods contaminated by $^{137}$Cs after the Chernobyl accident, *Health Phys.* **70**, 297–317.
24. Karaoglou, A., Desmet, G., Kelly, G.N. and Menzel, H.G. (1996) The Radiological Consequences of the Chernobyl Accident, European Commission, Luxembourg, EUR-16544.
25. European Commission, International Atomic Energy Agency, and World Health Organization (1996) One Decade after Chernobyl: Summing Up the Consequences of the Accident, International Atomic Energy Agency, Vienna.
26. Nuclear Energy Agency (1996) Chernobyl Ten Years On—Radiological and Health Impact, Organization for Economic Cooperation and Development, Paris.
27. Burkhart, W. and Kellerer, A.M. (1994) Radiation exposure in the Southern Urals, *Sci. Total Environ.* **142**, Nos. 1-2.
28. Degteva, M.O. (1996) Environmental dose reconstruction for the Urals population, in *Environmental Dose Reconstruction and Risk Implications,* National Council on Radiation Protection and Measurements, Bethesda, pp. 155–169.
29. Degteva, M.O., Kozheurov, V.P., Burmistrov, D.S., Vorobyova, M.I., Valchuk, V.V., Bougrov, N.G. and Shishkina, H.A. (1996) An approach to dose reconstruction for the Urals population, *Health Phys.* **71**, 71–76.
30. Degteva, M.O., Kozheurov, V.P., Vorobiova, M.I., Burmistrov, D.S., Khokhryakov, V.V., Suslova, K.G., Anspaugh, L.R., Napier, B.A. and Bouville, A. (1997) Population exposure dose reconstruction for the Urals Region, in *Assessing Health and Environmental Risks from Long-Term Radiation Contamination in Chelyabinsk, Russia,* American Association for the Advancement of Science, Washington, pp. 21–33.
31. Straume, T., Anspaugh, L.R., Haskell, E.H., Lucas, J.N., Marchetti, A.A., Likhtarev, I.A., Chumak, V.V., Romanyukha, A.A., Khrouch, V.T., Gavrilin, Yu.I. and Minenko, V.F. (1997) Emerging technological bases for retrospective dosimetry, *Stem Cells* **15**(Suppl. 2), 183–193.
32. National Research Council (1995) Radiation Dose Reconstruction for Epidemiologic Uses, National Academy Press, Washington.
33. Kozheurov, V.P. (1994) SICH-9.1—a unique whole-body counting system for measuring Sr-90 via bremsstrahlung. The main results from a long-term investigation of the Techa River population, *Sci. Total Environ.* **142**, 37–48.
34. Lucas, J.N., Hill, F., Burk, C., Fester, T. and Straume, T. (1995) Dose-response curve for chromosome translocations measured in human lymphocytes exposed to $^{60}$Co gamma rays, *Health Phys.* **68**, 761–765.
35. Haskell, E.H., Kenner, G.H. and Hayes, R.B. (1995) Electron paramagnetic resonance dosimetry of dentine following removal of organic material, *Health Phys.* **68**, 579–584.
36. Gilbert, E.S., Fix, J.J. and Baumgartner, W.V. (1996) An approach to evaluating bias and uncertainty in estimates of external dose obtained from personal dosimeters, *Health Phys.* **70**, 336–345.
37. Maruyama, T., Kumamoto, Y., Ichikawa, Y., Nagatomo, T., Hoshi, M., Haskell, E. and Kaipa, P. (1987) Thermoluminescence measurements of gamma rays, in [6], Vol. 1, pp. 143–184.
38. Haskell, E.H., Bailiff, I.K., Kenner, G.H., Kaipa, P.L. and Wrenn, M.E. (1994) Thermoluminescence measurements of gamma-ray doses attributable to fallout from the Nevada Test Site using building bricks as natural dosimeters, *Health Phys.* **66**, 380–391.
39. Bougrov, N.G., Vlasov, V.K., Kiryukhin, O.V. and Fatkulbayanova, N.L. (1995) Thermoluminescence measurements of ceramic samples from accidentally polluted territory of Southern Urals, *Radiat. Meas.* **24**, 493–498.

40. Bougrov, N.G., Göksu, H.Y., Haskell, E., Degteva, M.O., Meckbach, R. and Jacob, P. (1998) Issues in the reconstruction of environmental doses on the basis of thermoluminescence measurements in the Techa Riverside, *Health Phys.* **75**, 574–583.
41. U.N. Scientific Committee on the Effects of Atomic Radiation. (1962) Report of the United Nations Scientific Committee on the Effects of Atomic Radiation, United Nations, New York, General Assembly, Official Records: Seventeenth Session, Supplement No. 16 (A/5216).
42. Beck, H.L. (1980) Exposure Rate Conversion Factors for Radionuclides Deposited on the Ground, U.S. Department of Energy Environmental Measurements Laboratory, New York, EML–378.
43. Eckerman, K.F. and Ryman, J.C. (1993) External Exposure to Radionuclides in Air, Water, and Soil. Federal Guidance Report No. 12. Environmental Protection Agency, Washington, EPA 402-R-93-081.
44. Hicks, H.G. (1990) Additional calculations of radionuclide production following nuclear explosions and Pu isotopic ratios for Nevada Test Site events, *Health Phys.* **59**, 515–523.
45. Beck, H.L. and Anspaugh, L.R. (1991) Development of the County Database: Estimates of Exposure Rates and Times of Arrival of Fallout in the ORERP Phase-II Area. Comparison with Cumulative Deposition-Density Estimates Bases on Analyses of Retrospective and Historical Soil Samples, U.S. Department of Energy Nevada Operations Office, Las Vegas, DOE/NV–320.
46. Straume, T., Marchetti, A.A., Anspaugh, L.R., Khrouch, V.T., Gavrilin, Yu.I., Shinkarev, S.M., Drozdovitch, V.V., Ulanovsky, A.V., Korneev, S.V., Brekeshev, M.K., Leonov, E.S., Voigt, G., Panchenko, S.V. and Minenko, V.F. (1996) The feasibility of using $^{129}$I to reconstruct $^{131}$I deposition from the Chernobyl reactor accident, *Health Phys.* **71**, 733–740.
47. Straume, T., Marchetti, A.A. and McAninch, J.E. (1996) New analytical capability may provide solution to the neutron dosimetry problem in Hiroshima, *Radiat. Prot. Dosim.* **67**, 568–582.
48. WHO (1983) Environmental Health Criteria 25. Selected Radionuclides: Tritium, Carbon-14, Krypton-85, Strontium-90, Iodine, Caesium-137, Radon, Plutonium, World Health Organization, Geneva.
49. IAEA (1985) The Radiological Impact of Radionuclides Dispersed on a Regional and Global Scale: Methods for Assessment and their Application, International Atomic Energy Agency, Vienna, Technical Reports Series No. 250.
50. Gordeev, K.I., Kiselev, V.I., Lebedev, A.N. and Savkin, M.N. (1994) Method of retrospective reconstruction of radiological situation determining the internal irradiation upon the trace of nuclear tests, *Bull. Semipalatinsk Test Site/Altai* **No. 1**, 57–96 (in Russian).
51. Gordeev, K.I., Kiselev, V.I., Lebedev, A.N., Savkin, M.N. and Shoikhet, Y.N. (1995) Scientific grounds of models for calculating the doses of external and internal exposure for the population living on the radioactive vestige of nuclear explosion and some recommendations on using these models, *Bull. Semipalatinsk Test Site/Altai* **No. 1**, 56–79 (in Russian).
52. Shoikhet, Y.N., Kiselev, V.I., Loborev, V.M., Sudakov, V.V., Algazin, A.I., Lagutin, A.A., Zaitsev, E.V., Kolyado, I.B., Zelenov, V.I., Gabbasov, M.N. and Goncharov, A.I. (1999) Nuclear Tests at the Semipalatinsk Test Site. Radiation Impact on the Altai Region Population, Institute of Regional Medico-Ecological Problems, Barnaul, Russia.
53. Cederwall, R.T. and Peterson, K.R. (1990) Meteorological modeling of arrival and deposition of fallout at intermediate distances downwind of the Nevada Test Site, *Health Phys.* **59**, 593–601.
54. Takada, J., Hoshi, M., Endo, S., Yamamoto, M., Nagatomo, T., Gusev, B.I., Rozenson, R.I., Apsalikov, K.N. and Tchaijunusova, N.J. (1996) Thermoluminescence dosimetry of gamma rays from the fallout of the Semipalatinsk nuclear tests, in *Effects of Low-Level Radiation for Residents near Semipalatinsk Nuclear Test Site,* Research Institute for Radiation Biology and Medicine, Hiroshima University, Hiroshima, Japan, pp. 195–199.
55. Hicks, H.G. (1981) Radiochemical Data Collected on Events from which Radioactivity Escaped Beyond the Borders of the Nevada Test Range Complex, Lawrence Livermore National Laboratory, Livermore, CA, UCRL-52934.
56. Hicks, H.G. (1982) Calculation of the concentration of any radionuclide deposited on the ground by offsite fallout from a nuclear detonation, *Health Phys.* **42**, 585–600.
57. Anspaugh, L.R. and Church, B.W. (1985) Historical estimates of external gamma exposure and collective external gamma exposure from testing at the Nevada Test Site. I. Test series through Hardtack II, 1958, Health *Phys.* **51**, 35–51.
58. Henderson, R.W. (1991) Approximation of the Decay of Fission and Activation Product Mixtures, Los Alamos National Laboratory, Los Alamos, NM, LA-11968-MS.
59. Kirchner, T.B., Whicker, F.W., Anspaugh, L.R. and Ng, Y.C. (1996) Estimating internal dose due to ingestion of radionuclides from Nevada Test Site fallout, *Health Phys.* **71**, 487–501.

60. International Commission on Radiological Protection (1998) The ICRP Database of Dose Coefficients: Workers and Members of the Public, Elsevier Science, New York, ISBN 0 08 042 7510.
61. Whicker, F.W. and Kirchner, T.B. (1987) PATHWAY: A dynamic food-chain model to predict radionuclide ingestion after fallout deposition, *Health Phys.* **52**, 717–737.
62. Ward, G.M. and Whicker, F.W. (1990) Milk distribution and feeding practice data for the PATHWAY model, *Health Phys.* **59**, 637–643.
63. Whicker, F.W., Kirchner, T.B., Breshears, D.D. and Otis, M.D. (1990) Estimation of radionuclide ingestion: The "PATHWAY" food-chain model, *Health Phys.* **59**, 645–657.
64. Breshears, D.D., Kirchner, T.B., Otis, M.D. and Whicker, F.W. (1989) Uncertainty in predictions of fallout radionuclides in foods and of subsequent ingestion, *Health Phys.* **57**, 943–953.
65. Simon, S.L., Lloyd, R.D., Till, J.E., Hawthorne, H.A., Gren, D.C., Rallison, M.L. and Stevens, W. (1990) Development of a method to estimate thyroid dose from fallout radioiodine in a cohort study, *Health Phys.* **59**, 669–691.

# ESR INVESTIGATION OF ENVIRONMENTAL OBJECTS FROM THE SEMIPALATINSK NUCLEAR TEST SITE REGION

S.P. PIVOVAROV, A.B. RUKHIN, T.A. SEREDAVINA
*Institute of Nuclear Physics,*
*National Nuclear Centre of the Republic of Kazakhstan,*
*Almaty, 480082, Kazakhstan*

**Abstract**

It is discussed experimental data on study of the Semipalatinsk Nuclear Test Site soils by electron spin resonance method. The initial results and a perspective on the most effective directions for future investigations have considered. It is noted, that one should pay a necessary attention to this new approach in nuclear weapon tests consequences estimation.

## 1. Background

Among the numerous consequences of nuclear explosions, a radiation occupies the most important place and study of it influence results should be one of the basic problem for estimation the effects of nuclear weapons testing.

Despite the obvious reality of this problem, not enough attention has been paid until the present to this question. The basic interests of scientists were directed toward more simple problems, namely measuring radiating background and defining the remaining radionuclides concentrations. These works are certainly extremely important and we in no measure want to reduce their significance. We would like only to emphasize that these approaches do not actually provide any information on the influence on the environment of intensive, short-term radiation at the moment of explosion and not so intensive but long-term irradiation caused by decaying radionuclides.

It is well known that the large radiation doses can essentially modify many materials because of defects accumulation and can appreciably affect on kinetics of some redox processes, the destruction of organic compounds, the sorption and desorption processes and so on. Taking this into account, one can conclude, that for radioecological situation on nuclear test site study there is a significant interest in researching radiation defects in the soil and other environmental objects. This interest especially focuses on those objects that have received the most powerful radiation and thermal impacts from the explosions and on those ones that have been subsequently irradiated over a long period by the radionuclides remaining after the explosion.



*S.S. Hecker et al. (eds.), Nuclear Physical Methods in Radioecological Investigations of Nuclear Test Sites,* 173-179.
© 2000 *Kluwer Academic Publishers. Printed in the Netherlands.*

## 2. Electronic Spin Resonance method data

The most suitable method for achieving such goals is electronic spin resonance (ESR) which is one of the most powerful methods for studying the influence of radiation on condensed matter. In particularly, ESR makes it possible to provide with rather high sensitivity detailed information on a quantity and structure of radiation defects in various fractions of the soil. Of course, for complete information about the complex physical-chemical processes that had occured in soil as a result of radiation influence, the data achieved by only one method are obvious insufficient. Indeed, it is necessary to have also an information about nonparamagnetic products, arising as a result of containing nonpaired electron paramagnetic centre recombination and other reactions stimulated by free radicals and by radiation in general. For these purposes apparently a complex of various methods, first of all, chromatomassspectroscopy, will be necessary. However, at the first stage, one should carry out a reconnoitering research by ESR, which is a highly productive and direct method.

The initial ESR results should provide a general information on the radiation damage of soils, their fractions, and on similar effects for other objects of the environment. After some adaptation and methodical development, the determine results on the dose of irradiation on a given object value and the characterization of this irradiation type can be certain estimate. Finally, this preliminary stage will make it possible to choose the most interesting objects (or their fractions) for more detailed research by using another methods, such as chromatomassspectroscopy.

One of the ultimate aim of this research should be to reach conclusions about the ecological, genetic, medical, and other dangers connected with the intensive radiating influence of the nuclear weapons tests on soils, minerals, etc. in nuclear regions of the world. These potential dangers can be caused by the very high chemical activity of formed methastable radiation defects; however, as mentioned above, this question still remains practically unexplored.

The first results of researching the Semipalatinsk Nuclear Test Site (SNTS) soils by an ESR method were reported by us in Munich [1] and then in Hong Kong [2]. Thereafter, the research was much extended and the technical maintenance improved due to the grant of the Government of Japan - the powerful complex of modern ESR equipment, the main part of which is ESP 300E spectrometer produced by "BRUKER" firm (Germany).

Figure 1 presents an example of an electronic resonance spectrum of SNTS soils. This signal has a complex nature and consists of a wide signal of the ferromagnetic resonance as well as more narrow signals of the paramagnetic resonance of transient element ions and radiation defects. The shape and intensity of these signals vary from sample to sample and specify a presence in researched samples in different proportions of magnetic minerals, paramagnetic ions and methastable states with unpaired electrons. Such magnetic minerals are, for ex., strongly ferromagnetic magnetite $Fe_3O_4$ with g-factor of a central part about 9.0; weak magnetic hematite $Fe_2O_3$ with g-factor about 6.0 and mainly antiferromagnetic limonyte $FeOOH \bullet n\ H_2O$ with g-factor about 3.0. It is completely obvious that these minerals have a natural origin and we could not observe an appreciable correlation between the intensity of these signals and radionuclides contents (such a correlation one should expect, because of different sorption ability some of these minerals). However, this conclusion should be considered to be only preliminary. For an exact conclusion, the more careful division of the sample into mineralogical fractions and more extensive statistics are necessary. At the present time, rather significant work on study samples, carefully divided between magnetic and nonmagnetic, and also granulometrical fractions have been done, however, for a reliable substantiation of final conclusions, this work should be continued.

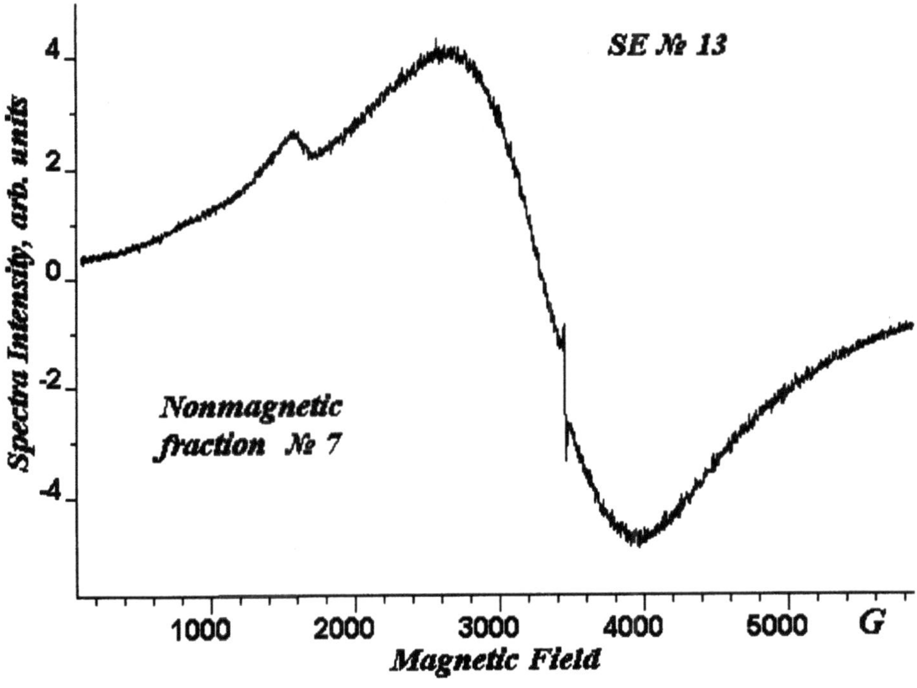


*Figure 1.* Typical ESR Spectrum of SNTS soil in wide range of magnetic fields

The paramagnetic resonance signal observed in the area, with g-factor ~4.3, is caused probably by an ion $Fe^{3+}$ in a orthorombic crystal field. The most suitable model of this ion is $FeO_2^{2-}$; however, the intensity of this signal also (from available data) does not noticeable correlate with the radionuclides contents.

In some samples, a characteristic $Mn^{++}$ signal that is present in soil, as a rule, in a very small amount (less ppm) is displayed. This signal as well as other paramagnetic ion signals visible on ESR spectra are natural impurities that do not require a special study.

The radiation origin of a narrow signal in around g=2.0 was proved by special experiments with additional artificial irradiation. These experiments have shown the complete identity of the signals registered in SNTS soils in Experimental Field region or on a trace of a radioactive cloud. These experiments also had demonstrated, that identical signals were observed in similar soils but taken far from Experimental Field and then specially irradiated.

The intensity of this signal is the greatest in the epicentre vicinities and then gradually very nonmonotonously reduced along a south-east trace. The number of electron spins giving this signal far exceed a calculated maximum quantity of unpaired electrons, which can make 1KBk of plutonium for 100 years. Hence, the basic part of a signal was generated by external irradiation and by short-living products. Research on the dependence of the shape of a signal from microwave power (Fig. 2) and research on the Q-band (Fig. 3) have shown that the signal has a rather complex nature and represents a superposition of several different paramagnetic centre signals. The registration of spectra at a very low level of microwave power has made it possible to allocate a specific signal, caused, apparently, with $E_1^1$ - centre in $SiO_2$. At a larger microwave power level, a signal of peroxy-centre [3, 4] is obviously allocated (Fig. 2), but these conclusions demand additional confirmation.

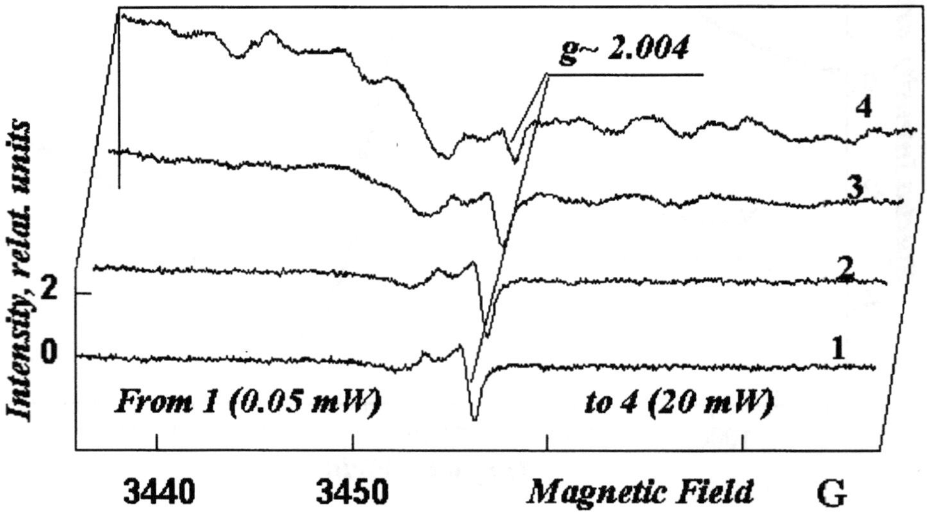


*Figure 2*. ESR signal shape (g~2.0) at various levels of microwave power.

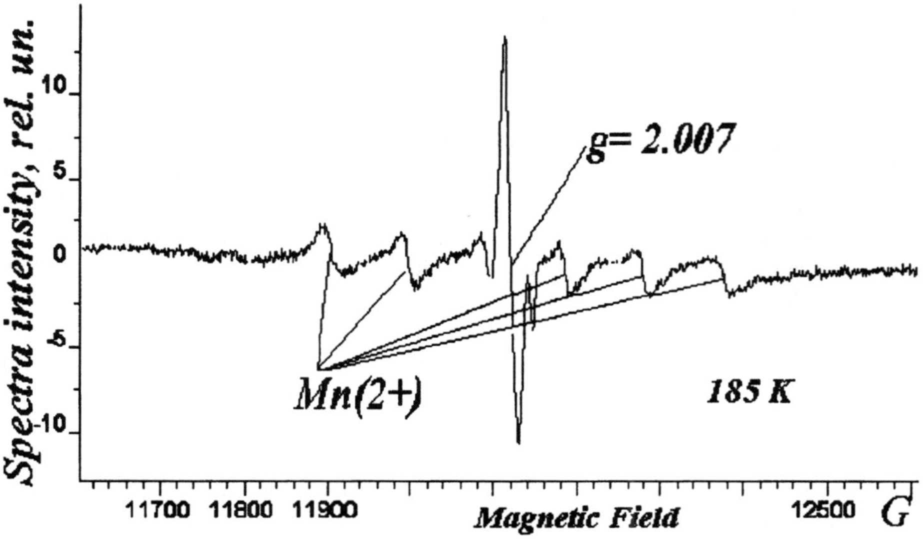


*Figure 3.* Radiogenic ESR signal registered in the Q-band.

Studying the distribution of a signal on mineralogical fractions has shown that, as should be expected, different fractions accumulate it with different efficiency (Fig. 4).

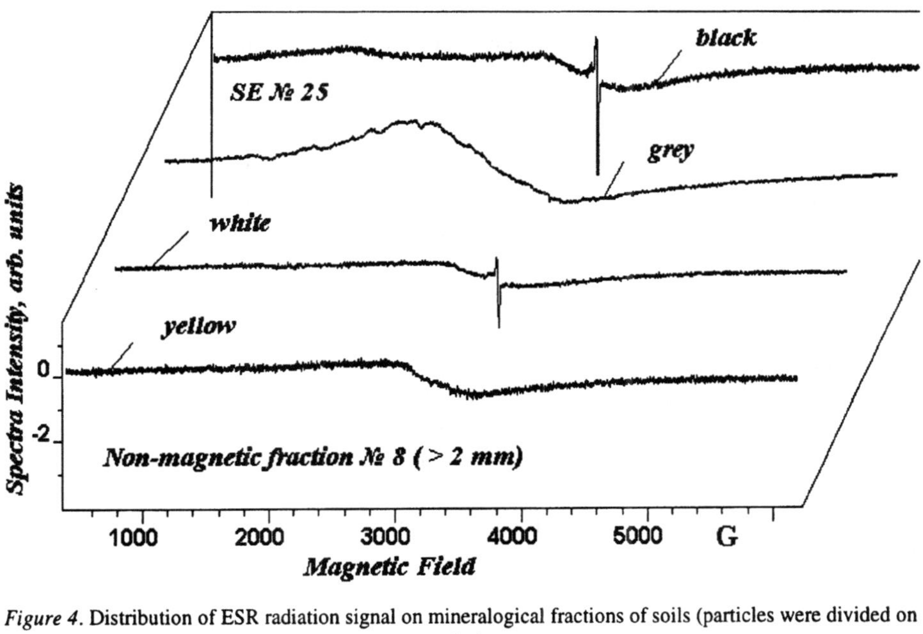


*Figure 4.* Distribution of ESR radiation signal on mineralogical fractions of soils (particles were divided on color).

These first results allow us to make some preliminary conclusions and to designate directions and prospects of further research.

The first, a radiation signal can be used to define a radiation dose distribution on the polygon territory and along traces of radioactive clouds. For this purpose it is necessary to find the mineral most sensitive to radiation and over a period of tens years keep this information about the irradiation. It is also necessary to allocate this mineral in the pure state in required amounts (share of grams) and to determine a constant of stabilized unpaired electrons radiation yield. After completing these steps, determining an integrated radiation dose will be a routine problem, limited only by the operations of the samples collecting, the allocation of a required fraction and ESR signal registration. Another methodical approach can be a series of additional artificial irradiations on *brutto* samples and extrapolation from a zero dose, as it was recommended in [3]. But in such a case, it is necessary to take into account the variations of a thermal stability of the defects in different fractions.

The second, there is a basis for the belief that the study of peculiarities of a radiation signal will make it possible to divide the effects of photon irradiation and particle irradiations because these kinds of rays have essentially differing linear energy transfer (LET) values. Therefore the local concentration of formed defects will be much higher then average one for beams with high LET and approximately equal to the average for rays with low LET. Because these distinctions in local concentrations usually result as distinctions in ESR relaxation characteristics, various methods of allocating these effects had been developed [5]. For a stricter substantiation of such experiments, it is necessary to use additional artificial irradiations with various types of rays and to develop an algorithm of the quantitative division of the appropriate effects.

The third, it is quite necessary to clear out the possible directions of the physical-chemical activity of radiation-induced paramagnetic centres, giving a narrow complex radiation signal with g-factor around 2.0. As it was noted above, these paramagnetic centres is really methastable states, kept in a firm matrix for relatively long times and showing, under certain conditions, a very high reactionability. In essence, these centres are free radicals, ion-radicals, threeplete states, etc. isolated from external environment.

A complete analogy to natural radioactivity is available here. Similar methastable radiogenic structures are also rather typical in nature, as well as natural radionuclides. But in this case, which is quite similar to an example with radionuclides, the background of these structures is more higher and among these structures one can find also products absent in natural conditions (analogy with Am, Pu, $Cs^{137}$ et al.)

The potential danger of such structures is that under appropriate conditions they can initiate for ex cancerogenic and mutagenic biochemical reactions. In addition, they can essentially influence on redox and other free radical processes occurring in living organisms.

On the whole this problem has remained completely unexplored until now. Only a few published articles indicate the real influence of such structures on minerals and on processes of ore flotation [6,7].

It is necessary to conduct further systematic work both to discover the quantity, structure, distribution, and basic directions of reaction ability of specific methastable structures found in the soils and minerals of SNTS and to estimate a dose load of various environmental objects.

## 3. References


1. Pivovarov S., Mosienko T. (1995) "ESR of Soils near a nuclear explosions," Proc. Int. Symposium ESR-Dosimetry, Germany, Munich, p.15.
2. Pivovarov S.P., Kazachevskii I.V., Seredavina T.A., Solodukhin V.P., Stakhov O.V. Anomal Intensity of ESR- Signal from glassed soil particles of Semipalatinsk nuclear test site epicenter (1997) Abstr. of the 1st Asia-Pacific EPR/ESR Symposium, Hong-Kong, p.106.
3. Ikeya M. (1992) New Applications of Electron Spin Resonance. "World Scientific" Singapore- New Jersey- London- Hong Kong, 499 p.
4. Matyash I.V., Brick A.B. et al. (1987) Radiospectroscopy of Quartz. Kiev. "Naukova Dumka", 166p. (in Russian).
5. Lebedev Ya. S, Muromtzev V.I. (1972) EPR and stabilised radical relaxation. Moscow "Chimiya" 254 p. (in Russian).
6. Gromov V.V. (1976) The Influence of ionising radiation on solids solution kinetics. Moscow "Atomizdat" 125 p. (in Russian).
7. Komov I.L. (1982) Radiation Mineralogy. Moscow "Energoizdat" 174 p. (in Russian).


## 4. Acknowledgement


The authors express their gratitude to the Government of Japan for providing the ESR equipment for carrying out of ecological research at the SNTS. They also thankful to ISTC. The main part of the work had been fulfilled within the framework of ISTC Project K-053.

# MONITORING THE MARSHALL ISLANDS FOR RESIDUAL WEAPONS TESTING FALLOUT: *TECHNIQUES AND FINDINGS*

S. L. SIMON
*Board on Radiation Effects Research, National Academy of Sciences,*[a]
*National Research Council, Washington, DC*
*Formerly of the Marshall Islands Nationwide Radiological Study, Majuro, Marshall Islands*

## Abstract

The Republic of the Marshall Islands, site of 66 nuclear weapons tests conducted by the U.S. Government, was monitored for residual radiological contamination from 1990 through 1994 by a study commissioned by the Marshall Islands Government. Several monitoring methods were used including *in situ* gamma spectrometry, soil sampling and laboratory measurement of gamma and alpha emitters, comparison of ground level measurements with aerial measurements (for a limit number of islands) taken 15 years earlier, and measurements of radionuclide uptake into edible plants and into plants used for medicinal purposes. The various techniques were used to assess present day exposure rates and possible future exposures of island inhabitants where remediation options are being considered. In addition, these data allowed various questions to be investigated including fractionation between plutonium and caesium as a function of distance from the test site. A brief summary of these various monitoring methods and related findings are presented.

## 1. Introduction

The United States Atomic Testing Program in the Pacific was conducted from 1946 through 1958 almost entirely in the Marshall Islands. The testing programme included detonation of 66 nuclear devices with a total explosive yield of 1.08 x $10^5$ kt TNT (DOE 1994; Simon and Robison 1997). Eighteen of these tests exceeded 1 MT in yield and contributed approximately 95% of the total yield. Using available information on nuclear testing in the Northern Hemisphere, one can estimate that approximately one-third of the global fallout inventory in the hemisphere originated from U.S. nuclear tests conducted in the Marshall Islands (Simon *et al.* 1999).

Though various monitoring programmes of the test site atolls and the atolls near the test sites had been conducted in the 50 years since the testing programme began, the

[a] Note new address in List of Authors.



*S.S. Hecker et al. (eds.), Nuclear Physical Methods in Radioecological Investigations of Nuclear Test Sites,* 181-189.
© 2000 *Kluwer Academic Publishers. Printed in the Netherlands.*

entire Marshall Islands had never been systematically monitored for residual radioactivity prior to 1990. For the five-year period, 1990 through 1994, the Republic of the Marshall Islands (RMI) Government undertook a radiological study of its 29 atolls to assess the radiological conditions at locations nationwide. Detailed findings of the Nationwide Radiological Study (NWRS) are reported in the open literature (see Graham and Simon 1996, Simon and Graham 1996, Simon and Graham 1997, Simon *et al.* 1999) and in a series of reports.* It is the purpose of this paper to summarise methods and findings of selected data that may be useful in studies of the nuclear test site at Semipalatinsk in the Republic of Kazakhstan and elsewhere.

## 2. Methods

The RMI is located in Pacific equatorial waters (see Simon and Graham 1996 for a map of the islands) and is composed of 29 atolls including more than 1200 islets. Although the total land area of the nations is a scant 180 $km^2$, the atolls are distributed over $6 \times 10^5$ $km^2$ of ocean. Consequently, logistics and choice of instrumentation were main considerations in designing the study, in addition to the considerations of cultural and language issues.

The primary method of the monitoring programme was the use of *in situ* gamma spectrometry measurements made on the larger islands (>400) supplemented by laboratory analysis of a variety of types of samples. Sampling sites were generally selected at random. However, at atolls where historical data indicated significant contamination, and where public and government interest was greater, systematic sampling (square grids on 200 m centres) was used to ensure uniform geographic coverage of the islands. In that case, *in situ* measurement sites were chosen near to the centre point of each grid square, except as necessary to avoid shorelines, disturbed areas, or man-made structures. The overall average spatial density of *in situ* gamma spectrometry measurements in the NWRS was 10 per $km^2$.

Soil profiles sampled in 5-cm increments to a total depth of 30 cm were also an important part of the survey. Characterisation of the vertical profile of $^{137}Cs$ activity is a parameter of considerable importance to estimating the aerial inventory. Over 200 soil profiles were acquired during the survey of the Marshall Islands. Generally, a ratio of 1 soil profile to each 6 *in situ* gamma measurements was maintained.

*In situ* gamma spectrometry measurements were made with high purity germanium detectors (HPGe) of 40% nominal efficiency (relative to a 3" x 3" NaI detector). These detectors were attached to 7-l liquid nitrogen cryostats, which could maintain suitably low temperatures in a tropical environment for over 3 days. The minimum detectable *in situ* count rate for $^{137}Cs$ was estimated to be 0.0085 c/s for a counting time of 2 hours. That count rate corresponds approximately to 15 Bq/$m^2$ of $^{137}Cs$. *In situ* detection limits for $^{241}Am$ and $^{60}Co$ were determined to be approximately 100 and 10 Bq/$m^2$, respectively.

Laboratory measurements for gamma emitters were conducted with two electro-cooled HPGe detectors of 40% efficiency with extended low-energy response. Detectors

---

* Available at http://radefx.bcm.tmc.edu/marshall_islands/.

were each housed in 1 inch thick lead shields located in an air-conditioned building of wood construction. The counting facility was built on a bed of crushed coral that was dredged from the lagoon on Majuro Atoll. The building was surrounded on 3 sides by the lagoon, approximately 4 meters from the building, thus ensuring a low background environment. The minimum detectable concentrations for $^{241}$Am, $^{137}$Cs, and $^{60}$Co were estimated to be 2.0, 0.3, and 0.2 Bq/kg, respectively, for a 12 h counting period.

All measurements were conducted in a radiological laboratory built in Majuro, the capital city of the Marshall Islands, specifically for the purposes of this study. The study was extensively assisted in the field and in the laboratory by Marshallese citizens trained by the laboratory staff.

Native foods including coconuts and other tropical fruits were collected as well as more than 800 surface soil samples. The fruits, soil profiles, and surface soil samples were analysed for gamma emitters with an emphasis on determining concentrations of $^{137}$Cs; the surface soil samples were also analysed for $^{239+240}$Pu.

Plutonium content was determined by alpha spectrometry after chemical extraction from the soil. Extraction was based on complete soil digestion, extraction with anion exchange columns, followed by microprecipitation and plutonium mounting with neodymium fluoride. To maintain approximately equal counting times (on the alpha spectrometer) necessary to attain a measurement precision of $\pm$ 10% at the 1$\sigma$ confidence level, the soil mass was adjusted for each sample based on a prior gamma spectrometry measurement of the $^{241}$Am.

## 3. Findings

### 3.1. DETERMINATION OF REGIONAL BACKGROUND

Estimates of global fallout in the mid-Pacific were obtained from published data and decay findings corrected to 1994. Data of $^{90}$Sr deposition were used to derive caesium estimates by assuming a production ratio of $^{137}$Cs/$^{90}$Sr of 1.6. The data on global fallout were examined with respect to latitudinal and precipitation variation. Generally, global fallout deposition increases with increasing latitude in the Northern Hemisphere, though it is also a strong function of the annual precipitation rate. Within the Marshall Islands archipelago, a strong north-south rainfall gradient exists with annual precipitation of 300 cm typical in the southern atolls and 100 cm in the northern atolls. Thus, the possibility of higher deposition in the northern atolls was likely offset by the lower rainfall there. The expected contribution of $^{137}$Cs from global fallout on the atolls of the Marshall Islands was estimated to be relatively constant because of this opposing effect—approximately 400 and 800 Bq/m$^2$.

The activity ratio of $^{137}$Cs to $^{239+240}$Pu from global fallout was investigated in the early 1970s by the U.S. Health and Safety Laboratory and reported to be a constant value in the North Temperate Zone (Beck and Krey 1983). Decay correcting those data to 1994 gave a ratio of caesium to plutonium of about 38.4 $\pm$ 0.36. Thus, the best estimate of plutonium in the environment of the Marshall Islands from global fallout is between 11 and 22 Bq/m$^2$. Assuming unit density for the top 5 cm of soil and a uniform distribution within this layer, the contribution of plutonium from global fallout was estimated to be between 0.2 and 0.4 Bq/kg.

### 3.2. COMPARISON OF SURVEY METHODS

Exposure-rate estimates (μR/h) and aerial inventory of $^{137}$Cs (Bq/m$^2$) from the NWRS were compared (Simon and Graham 1998) with data from an aerial survey sponsored by the U.S. Department of Energy in 1978, using helicopter-mounted NaI detectors (Tipton and Meibaum 1981). The data originated from measurements made at 11 atolls and 2 separate reef islands. The total number of individual islets for which data were compared totalled 143. In the comparison, 728 ground-level *in situ* spectrometry measurements from the NWRS were matched with the aerial survey data collected by Edgerton, Germeshausen, and Grier, Inc. (EG&G) some 14 years earlier.

The data sets were compared using graphical and statistical techniques including scatterplots, and probability plots. A primary observation from scatterplots indicated that there was not a perfectly linear relationship between the EG&G and NWRS data sets. Both data sets agreed best in the upper range of the variables—in the ranges from 1 to 50 μR/h for exposure-rate and from $10^4$ to 5 x $10^5$ Bq/m$^2$ for aerial inventory. Below these values, the EG&G data consistently had higher estimates as compared to the NWRS data. For example, EG&G aerial inventory data were about 2.5 times greater at the 50th percentile and about 10 times greater at the 1st percentile (the lowest values).

The differences in findings were due to a variety of differences in calculation methodology and assumptions—particularly with respect to relaxation length—and detection limits for the two techniques. In the NWRS, the minimum detectable *in situ* count-rate for $^{137}$Cs was estimated to be about 1 nR/h or 15 Bq/m$^2$ of $^{137}$Cs for a counting time of 2 h (Simon and Graham 1997) compared to about 0.25 μR/h or 3000 Bq/m$^2$ for the aerial survey technique (in 1978). Of the locations compared in this report, 15% were estimated by the NWRS to be <1000 Bq/m$^2$. Soil inventory estimates by the NWRS extended down to about 150 Bq/m$^2$. The *in situ* measurement technique has a high degree of sensitivity relative to the aerial survey due to the long counting times that are possible. Consequently, many data points reported by the NWRS are significantly lower than the corresponding aerial survey values. Data from the two monitoring programmes, however, yielded exceptionally good agreement considering the number of practical differences in methodology and technology.

### 3.3. *IN SITU* DETECTOR CALIBRATIONS

Calibration of the *in situ* spectrometer was accomplished using calculations based on theory (see Beck 1972, ICRU 1994) and by comparing *in situ* measured count rates with the $^{137}$Cs inventory determined by laboratory measurement of soil profiles collected in the vicinity of the *in situ* measurement. Count rates and inventories correlated well over four magnitudes of range. A relationship between aerial inventory (Bq/m$^2$) and *in situ* count rate (c/s) was determined to be:

$$^{137}\text{Cs Inventory (Bq/m}^2) = 4090 \text{ x } (\text{c/s})^{1.04} \; (R^2 = 0.82).$$

Though we used a power-function to describe the relationship, the exponent on the count rate was close to unity, indicating that the relationship was nearly perfectly linear.

## 3.4. NATIONWIDE PATTERNS OF CONTAMINATION

A distinct pattern of decreasing soil inventory of $^{137}Cs$ with decreasing latitude was observed (see Simon and Graham 1997). This pattern was generally expected owing to the location of the test sites in the northern part of the nation and to the normal westerly tradewinds.

The maximum value of deposition at each atoll was interpreted to be closest to the original value of deposition (after accounting for radioactive decay). This assumption is most likely a good one for atolls that lie at distances of one hundred km or more from the test sites, though it is less certain for the test sites or nearby atolls. All islands in the Marshall Islands are coral and are virtually flat with highly porous soil with the result that precipitation is quickly absorbed into the soil. Standing water following storms is very rare; furthermore, there is no apparent erosion from runoff, which might lead to collection of radioactivity in localised areas. Such assumptions might not be valid at other sites worldwide due to variations in topography, which could lead to erosion and collection of radioactivity.

Atolls at latitudes of 9° N or less showed little evidence of having received local fallout deposition. The maximum observed inventories at those locations were 900 Bq/m$^2$ or less. Between 9° N and 11.5° N (where Bikini and Enewetak are located), the maximum observed aerial inventories on the atolls increased curvilinearly with increasing latitude to a maximum of about 8 x 10$^5$ Bq/m$^2$.

Data collected on plutonium concentrations in the upper soil horizon (top 5 cm) showed a similar pattern with latitude. At locations south of 9° N, the concentrations were 1 Bq/kg or less. Between 9° N and 11.5° N, the concentrations increased curvilinearly to about 5000 Bq/kg. The concentrations of $^{239+240}Pu$ in the analysed soil samples covered a range of nearly five orders of magnitude, from a low value of 0.042 Bq/kg at Taongi Atoll to 4140 Bq/kg at Enewetak Atoll.

## 3.5. RELATIONSHIPS BETWEEN RADIONUCLIDES

Concentrations of $^{241}Am$ were also measured in surface samples. Because $^{241}Am$ is a decay product of $^{241}Pu$, the inventory of americium in samples is usually correlated closely with the inventory of total plutonium or with $^{239}Pu$, the plutonium isotope of greatest abundance (and interest). Data on concentrations of $^{241}Am$ and $^{239}Pu$ were compared over a range of three orders of magnitude and were found to be significantly correlated. A relationship between concentrations of the two radionuclides was estimated by linear regression by fitting to a power function:

$$^{239}\text{Pu (Bq/kg)} = 3.8 \times (^{241}\text{Am})^{0.857} \; [R^2 = 0.65].$$

Measurement of $^{241}Am$ either *in situ* or in the laboratory is significantly easier than measurement of plutonium because of its relatively abundant gamma emission near 60 keV. Thus, estimates of $^{239}Pu$ can generally be made in areas contaminated with nuclear weapons residues by measurement of $^{241}Am$.

Relationships between other radionculides were also examined in samples of surface soil (0-5 cm). It was found, as expected, that most other interrelationships were less significant. Differences in chemical behaviour of each nuclide in the soil would lead to different degrees of permeation over the 45 years since the testing programme, thus reducing the correlation between pairs of nuclides. For example, plotting concentrations of $^{239}$Pu against $^{137}$Cs showed significant scatter. The following relationship was determined:

$$^{239}\text{Pu (Bq/kg)} = 5.5 \times (^{137}\text{Cs})^{0.54}\ [R^2 = 0.30].$$

The relationship between $^{155}$Eu and $^{137}$Cs was examined and found to be:

$$^{155}\text{Eu (Bq/kg)} = 1.3 \times (^{137}\text{Cs})^{0.43}\ [R^2 = 0.44].$$

## 3.6. FRACTIONATION EFFECTS

The ratio of $^{239+240}$Pu to $^{137}$Cs in the soil over a distance of 1000 km from the test sites was examined to determine if there were evidence for fractionation, i.e., differential deposition. It was found that the plutonium-caesium (Pu-Cs) ratio varied considerably with distance from the Bikini test site, generally decreasing with increasing distance.

Using estimates of the total aerial inventory derived from measurements of soil profiles, the Pu-Cs ratio data was fit to a function of distance using unweighted, non-linear least squares:

$$\text{Pu-Cs} = 1.85 \times e^{(-0.00483 \times \text{km})}\ [R^2 = 0.18, n=67].$$

Though the relationship was not strong, it was statistically significant, indicating that the ratio of Pu to Cs decreased approximately 133 times over the 1000 km distance, equivalent to a decrease of 38% per 100 km distance from the test site.

These observations indicate that plutonium as compared to caesium was depleted from the nuclear cloud at earlier times or at shorter distances. The reasons for the fractionation effect between plutonium and caesium are likely related to differences in their condensation temperatures as well as to differences in their chemical affinity for condensation nuclei. Both processes result in the likelihood of plutonium to form particles susceptible to more rapid gravitational settling.

## 3.7. RADIOACTIVITY IN PLANTS

The most commonly used food plant in the Marshallese culture and the most common type of tree is the coconut palm (*Cocos nucifera*). Coconuts are used at many different stages of growth and in a variety of prepared foods. Thus, coconut was the primary type of fruit sampled and analysed for radioactivity. Other fruits were obtained in the NWRS, though sample size was not equal among the atolls. These sample types included breadfruit (*Artocarpus altilis*), Pandanus (*Pandanus tectorius*), arrowroot (*Tacca leontopetaloides*), and lime (*Citrus aurantifolia*). Concentrations of $^{137}$Cs in coconuts varied greatly within the RMI depending on location. In general, concentrations were proportional to soil concentrations and ranged from <0.1 Bq/kg to >2300 Bq/kg in the coconut fluid and <5 Bq/kg to >21 000 Bq/kg in the coconut meat (dry).

The leaves or fruit of five species of plants used for purposes of traditional medicine were sampled at 14 of the 29 atolls of the Marshall Islands. The $^{137}$Cs concentrations were measured in 295 samples (Duffy *et al.* 1999). The concentration of $^{137}$Cs in each species generally decreased with increasing distance and decreasing latitude from the weapons testing sites in the northern region of the country. Of the species sampled, *Polypodium scolopendria* exhibited the highest mean $^{137}Cs$ concentration; varying from 2 to 11 times the concentrations of other species, depending on location. Concentrations in *Polypodium scolopendria* varied from 10 Bq/kg to 10 000 Bq/kg (dry wt.) Other species could not be statistically differentiated in terms of $^{137}Cs$ uptake.

### 3.8. DIET DESCRIPTIONS AND PROJECTED DOSES

The radiological measurement data in foods and the environment were used to estimate the expected effective dose rate in 1994, using an assumed set of lifestyle and dietary assumptions. Present-day dietary data were obtained by interview and reported by Dignan *et al.* (1994). Those data indicated that locally grown food contributed only about 18% of the total caloric intake for one particular community interviewed (Rongelap community). Total caloric intake rates were estimated to be 1900 $\pm$ 100 (1 SE, n=48) and 2750 $\pm$ 146 (1 SE, n=68) kcal/d for women and men, respectively. The applicability of the dietary day to other Marshallese communities was limited, however, because this group of people is receiving surplus USDA food and financial compensation. Nevertheless, the information on food types consumed and preferred could be extended with some assumptions to other Marshallese communities.

A primary assumption necessary to predict present-day doses was the proportion of food consumed that was grown locally. Various assumptions were made in consultation with the community members. One dietary model used for the internal dose calculations assumes that 75% of the dietary intake is from a mixture of locally grown food, the remainder being imported rice. The relative proportions of locally grown food were extrapolated from the diet of Dignan and included various local fruits (coconut, breadfruit, etc.)

Median effective doses at each atoll and a reasonable uncertainty range were estimated for residents of each atoll. Median doses from $^{137}Cs$ ranged from a low of 3 x $10^{-3}$ mSv/y at distant atolls to about 8 mSv/y near the test sites, with the range due to uncertainty yielding doses about 3 times higher and lower.

The effective dose from global fallout $^{137}Cs$ deposited in the equatorial Pacific region (400 to 800 Bq/m$^2$) is useful to provide a basis for comparison. The effective dose resulting from global fallout of that magnitude results in annual doses of about 6 x $10^{-3}$ mSv/y to about 0.1 mSv/y. Thus, the predicted doses at about ten of the atolls exceed the likely dose from global fallout.

Doses resulting from ingestion of preparations of medicinal plants varied, depending on location and amount of plant material consumed, but it generally contributed only a few percentage points, at most, to the total effective dose. By assuming one remedy was consumed yearly from each of the five plant types sampled, annual effective doses (mSv) using median concentration values were 4 x $10^{-4}$ in the southern-most islands, 6.1 x $10^{-3}$ at mid-latitude islands, and 1.2 x $10^{-2}$ in northern islands. Considerable uncertainty was acknowledged to exist regarding the methods of preparing medicinal mixtures, the amount of plant material used, and the frequency of consumption. It is clear that there is much to be learned about the practice of traditional medicine in the Marshall Islands, in particular, the frequency of use and the rates of consumption of plant-based remedies.

## 4. Discussion and Conclusions

On the basis of the various studies conducted as part of the NWRS, several conclusions concerning the radiological status of the Marshall Islands can be summarised. These include the following findings:

- Global fallout is estimated to be 400 to 800 Bq/m$^2$ for $^{137}$Cs and 0.5 to 2 Bq/kg for $^{239+240}$Pu in the mid-Pacific.
- Four southern atolls show no evidence of having received local fallout from the tests at Bikini and Enewetak.
- It cannot be conclusively determined whether ten of the atolls received local fallout or not.
- Explanations for the levels of $^{137}$Cs in some southern atolls being lower than global fallout estimates are: 1) weathering effects resulting in downward migration; 2) dilution of surface soil activity with clean humic material from litter fall or with coral material brought up from deep soil horizons by human or animals; or 3) erosion from ocean waves.
- Soil inventory of $^{137}$Cs is nearly constant among the atolls within the latitude range of 4° N to 9° N. The $^{137}$Cs in the soil was shown to increase sharply with increasing latitude between 9° and 11.5° N (where Bikini and Enewetak atolls are located).
- The sum of external and internal dose from $^{137}$Cs is predicted to vary from 3 μSv/yr to 10 mSv/yr, using a diet and lifestyle model determined by the survey.
- The residual radioactivity at the south to mid-latitude atolls (4° N to 10.5° N) does not pose any measurable health hazard today nor will it in the future.
- A number of islands in the atolls of Enewetak, Bikini, and Rongelap should have limited remediation before communities are encouraged to reinhabit those lands. This recommendation is contingent on the reasonable assumption that the Marshallese will continue to consume locally grown food.

The above conclusions drawn from studies of the NWRS are potentially useful to planning radiological studies at Semipalatinsk and other former nuclear testing sites. In particular, it is emphasised here that the contribution of global fallout must be determined, and locations within the area of interest systematically evaluated for evidence of having received contamination from the testing site. They should also be possibly ranked in order of degree of contamination, and evaluations of health hazards in the present and future should be distinguished from those in the past. In addition, realistic exposure scenarios should be constructed on the basis of knowledge of dietary habits and their variations among members of the local population.

## 5. Acknowledgements

The Nationwide Radiological Study was supported by the Government of the Republic of the Marshall Islands. Numerous people who assisted with data collection and sample analysis include James Graham, Susan Duffy, Andy Borchert, Andrew Barron, Sheila Como, and numerous Marshallese.

**Author's Note:** This manuscript does not represent the opinion of the National Academy of Sciences or the National Research Council.

## 6. References


1. Beck, H. L.; DeCampo, J. and Gogolak, C. 1972. *In situ* Ge(Li) and NaI(Tl) gamma-ray spectrometry for the measurement of environmental radiation. New York: Health and Safety Laboratory, U.S. Atomic Energy Commission, HASL-258.
2. Beck, H. L.; Krey, P. W. 1983. Radiation exposures in Utah from Nevada nuclear tests. Science 220(1 April): 18-24.
3. Dignan, C.; Calf, J.; Webb, D.; Mackerras, D.; Franke, B. 1994. Dietary survey of Rongelapese living on Mejatto Island. Rongelap Resettlement Project. Available from the Rongelap Local Government, Majuro, Marshall Islands.
4. Duffy, S., Simon, S. L., Whicker, F. W. 1999. "137-Cs contamination of plants used for traditional medicine and implications for human exposure." Journal of Environmental Radioactivity 46(1):27-44.
5. Graham, J. C.; Simon, S. L. 1996. "A Study of $^{137}$Cs in Soil Profiles from the Marshall Islands, Science and the Total Environment 183: 255-268.
6. ICRU. 1994. Gamma-Ray Spectrometry in the Environment. ICRU Report No. 53. Bethesda, MD: International Commission on Radiation Units and Measurements.
7. Simon, S. L.; Graham, J. C.; Borchert, A. 1999. "Concentrations and spatial distribution of plutonium in the terrestrial environment of the Republic of the Marshall Islands." The Science of the Total Environment 229(1-2): 21-39.
8. Simon, S. L.; Graham, J.C. 1996. "Dose Assessment Activities in the Republic of the Marshall Islands." Health Physics 71(4):438-456.
9. Simon, S. L.; Graham, J. C. 1997. Findings of the First Comprehensive Radiological Monitoring Program of the Republic of the Marshall Islands." Health Physics 73(1):66-85.
10. Simon, S. L.; Graham, J. C. 1998. "A comparison of aerial and ground level spectrometry measurements of $^{137}$Cs in the Marshall Islands." Environmental Monitoring and Assessment - An International Journal 53(2): 363-377.
11. Simon, S. L.; Robison, W. L. 1997. "A Compilation of Atomic Weapons' Test Detonation Data for U.S. Pacific Ocean tests." Health Physics 73(1):258-264.

## EXTERNAL EXPOSURE ESTIMATES FOR INDIVIDUALS NEAR THE NEVADA TEST SITE[a]

R. W. HENDERSON AND R. F. SMALE
*Environment, Safety, and Health Division,*
*Los Alamos National Laboratory, Los Alamos. NM 87545*

### Abstract

Individuals living near the Nevada Test Site were exposed to both β and γ radiations from fission products and activation products resulting from the atmospheric testing of nuclear devices. These exposures, and the resultant doses were functions of the amount of material deposited, the time of arrival of the debris, and the amount of shielding afforded by structures. Results are presented for each of nine generic life-styles. These are representative of the living patterns of people residing in the area. For each event at each location for which data exist. a representative of each life-style was closely followed for 30 d. The results of these detailed calculations were then extrapolated to the present, employing a stochastic model. Results displayed are the geometric means and standard deviations derived from 25 independent determinations of the various quantities shown. For each determination, required parameters were randomly selected from appropriate distributions. Calculations yielded estimates for: 1) whole-body and skin dose due to γ rays from material on the ground: 2) skin dose due to β particles from material deposited directly on the skin; and 3) skin dose due to β particles from material deposited on the ground. Organ dose estimates were established from the whole-body dose using appropriate conversion factors. For the homemaker life-style, the uterus dose was also calculated as a function of time for 9 mo. This assisted in estimating fetal dose as a function of gestation.

### 1. Introduction

External dose is defined in this paper as that dose to the whole body and various organs resulting from radionuclides external to the body. An external dose assessment model was designed to follow activities of an individual through a typical 30-d period following an above-ground nuclear event at the Nevada Test Site and use the results to

[a] This paper is reproduced from the *Health Physics* journal (Health Phys. 59: 715-721; 1990) with permission from the Health Physics Society.



*S.S. Hecker et al. (eds.), Nuclear Physical Methods in Radioecological Investigations of Nuclear Test Sites,* 191-202.
© 2000 *Kluwer Academic Publishers. Printed in the Netherlands.*

estimate the total dose received. The model was implemented using a stochastic code that made multiple, independent determinations of the various quantities reported. For each loop through the code a residence was selected and the normalized exposure rate at H + 12 h was estimated from the input. Cloud arrival time was also derived from input data for the location.

A life-style data table was entered to find the shielding situation for the individual at cloud arrival. Possible situations were: at home in the bedroom, at home in the living room, outdoors, in a public building, or in an automobile. A shielding factor for the situation was derived from the data tables, the duration of the situation was calculated, the unshielded exposure rate was derived and integrated to yield the unshielded exposure, the shielding factor was applied, and the shielded exposure for the period was derived. This process of deriving an unshielded exposure and a shielded exposure was repeated, and the results were accumulated for 30 d.

An "effective shielding factor" was calculated as the ratio of the shielded to the unshielded exposure for the 30-d period. The unshielded exposure from the end of the 30 d to the present was then calculated. The "effective shielding factor" was applied to this result, and the shielded exposure for the period from 30 d to the present was derived. This value was added to the previous accumulation and was the estimated exposure from the time of cloud passage to the present for the pass. This procedure was repeated 25 times for each life-style for each event at each location in the data base. All individual results were stored for statistical processing.

The stochastic feature of the model was provided by selecting values for the various quantities needed from appropriate distributions during multiple passes through the calculational parts of the code. The various input parameters and tabulated data, representing central values and appropriate measures of dispersion, were either read as data or assumed. Three separate distributions were used during execution of the code.

All exposure rate values were assumed to be distributed lognormally; all values for time, either cloud arrival or changes from one location to another, were assumed to be distributed normally. The distribution of homes for generic life-styles as assumed to be uniformly distributed over the 447 examples entered in the data tables. The individual shielding values for living areas and bedrooms in the data tables were distributed normally, and the standard deviation was included in the tables.

The model was implemented to yield dose estimates for non-specified individuals using generic inputs. Living pattern information was required for each generic group. These living patterns included the times spent in the various rooms of the house, and also the time spent outdoors, in public buildings, and in an automobile. Time data were specific for the seasons of the year and day of the week (weekday, Saturday, Sunday). These representative individual values could later be combined with population information to provide collective dose estimates.

## 2. Calculational Methods

Figure 1 displays a representation of four shielding situations considered. Shielding values for the bedroom and living room were taken from the table of 447 homes used as input. The value for the "other building" was estimated to be 0.5, with a standard deviation of 0.125. The model follows the movement of an individual as he or she moves from one shielding situation to another in a time-ordered sequence. Also included in the shielding situations was an automobile, since this was considered an important activity by some respondents in interviews. The value for the automobile was taken to be 0.8, with a standard deviation of 0.2.

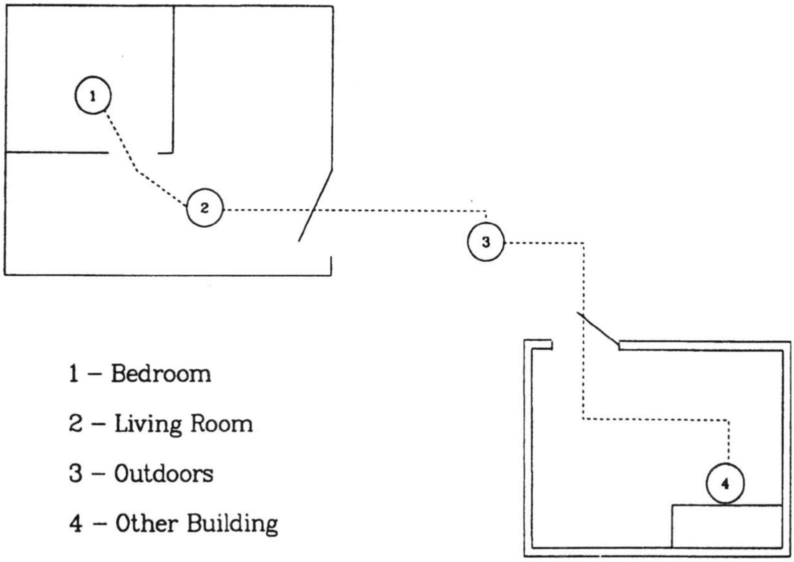


*Figure 1.* Representation of shielding situations used in the external dose model.

The model assumed that essentially the entire population was represented by eight life-styles. The first five life-styles were age-dependent: newborn—birth to 4 mo: infant—4 mo to 1 y; preschool—1-6 y; elementary student—6-12 y: and secondary student—12-19 y. The next three adult life-styles were occupation-dependent: homemaker—employed full-time: employed outdoors—farmer, telephone lineman, etc.; and employed indoors—school teacher, sales clerk, etc. A ninth life-style provided a unique category used to demonstrate the effects of shielding on the derived doses for the other life-styles: shepherd—full-time, outdoors.

Living patterns representing each of the first eight life-styles were derived from information received during the Allen vs. the United States trial (Civil Action No. C-79-5 15. U.S. District Court for the District of Utah Central Division, filed 30 August I979), the responses to the life-style survey done in the 10-county area near the Nevada

Test Site, personal experience, and interviews. The shepherd life-style is simply defined as being outdoors all of the time.

The derived living patterns were input to the code as tables containing the times that a member of the life-style moves from one location to the next. These times were assumed to be distributed normally about the value given in the table, with a standard deviation of 8.5 min. This resulted in a 2σ error of 15 min in the distributions. As the code followed the individual through the 30 d following the event, it used the life-style information for the individual to derive the shielding situation. This shielding situation was specific to the individual at that time. These data were not combined to estimate average residence times. The appropriate table was always entered to derive the shielding situation at the time in question.

Figure 2 displays the living patterns for the elementary student life-style. Figure 3 displays the living patterns for the homemaker life-style. In these figures, Friday is used to represent all weekdays.

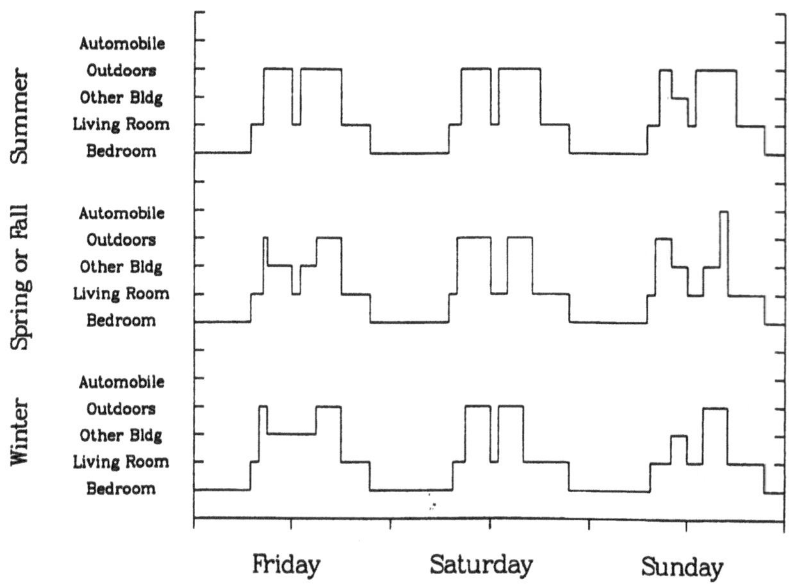


*Figure 2.* Living patterns for the elementary student life-style.

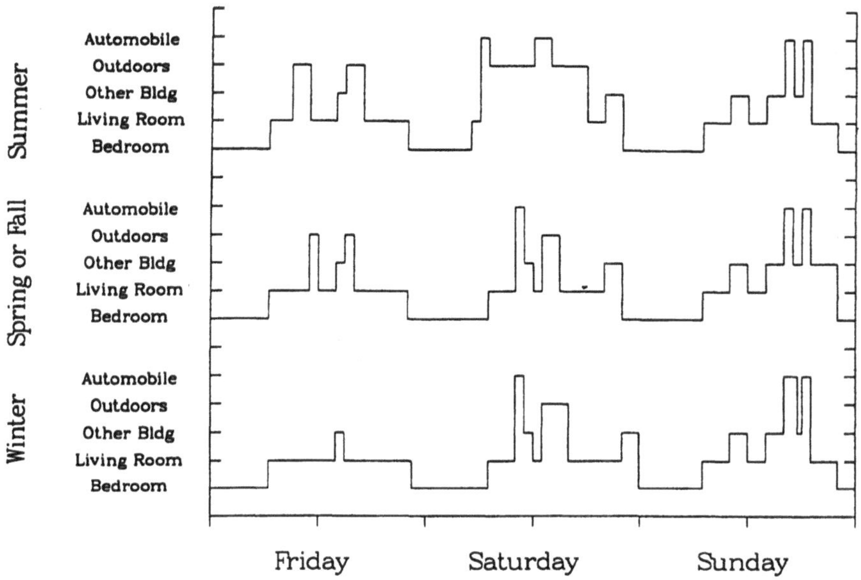


*Figure 3.* Living patterns for the homemaker life-style.

Exposure rate vs. time data are taken from the work of Hicks (1981). The exposure-rate data points given between 1 h and 50 y after detonation were approximated with sums of exponentials (typically 11), and this function was used for subsequent calculations.

It has always been desirable, if not necessary, to express the decay of the fallout radiation field with some sort of mathematical function. The first useful approximations were typically power functions of time. The power function, $T^{-1.2}$, was widely used in early work involving fission products. This was recognized as an approximation, and various attempts were made to try to improve it. Various schemes were used, including changing the exponent at various points in time. All of these functions could be made to fit the observed data very well over limited time ranges. Thus, the observed data were approximated not with a single power function, but with a series of power functions.

A different approximation scheme is suggested by the fact that the decay of the fission and activation products is the sum of a large number of exponential functions. The advantages of using sums of exponentials include the ability to approximate the given data to almost any degree of precision desired. The derived sums are all within 2% of the data points over the domain of the data. Further, the derived function is continuous and integration is straightforward. Sums of exponential functions have been derived to represent all of the events as described by Hicks (1981).

The Hicks report also gives the radionuclide composition of the fallout as a function of time. These data have been used to provide input parameters for Loevinger's equations ( Hine and Brownell 1956). The results of these calculations give β skin dose rate vs. time curves from material in the air and deposited on the skin during cloud passage and material deposited on the ground. The curves for the ground-deposited material are specific for a child or an adult. These four curves (air, skin, child-ground, and adult-ground) for each event have again been fit with sums of exponentials over 1 h to 1 mo. These functions are then used for integrations during periods of exposure.

The radionuclide data were also used to derive organ dose conversion factors. The data given by Kocher (1980) were used with the Hicks (1981) data to derive the ratio of specific organ dose to whole-body dose as a function of time for a typical event. These ratios were essentially constant from 1 h to 50 y and were used in the code as constants.

Figure 4 shows a stylized representation of an exposure-rate curve superimposed on the exposure modifiers for a 3-d period. An exposure modifier of 1.0 represents a period of time outdoors, a value of 0.5 represents time in a public building, the value below 0.5 represents a value for a bedroom, and the value slightly above 0.5 represents the living areas of a home. The code integrates under the curve between each change in shielding situation and applies the exposure modifier to the integrated exposure:

$$E_i = EM_i \int_{t_i}^{t_{i+1}} ER(t)dt$$

$$E_{30d} = \sum_{all\ i} E_i, \tag{1}$$

where $ER(t)$ is the exposure rate function, $EM_i$ is the exposure modifier for the period. And $E_i$ is the exposure for the period.

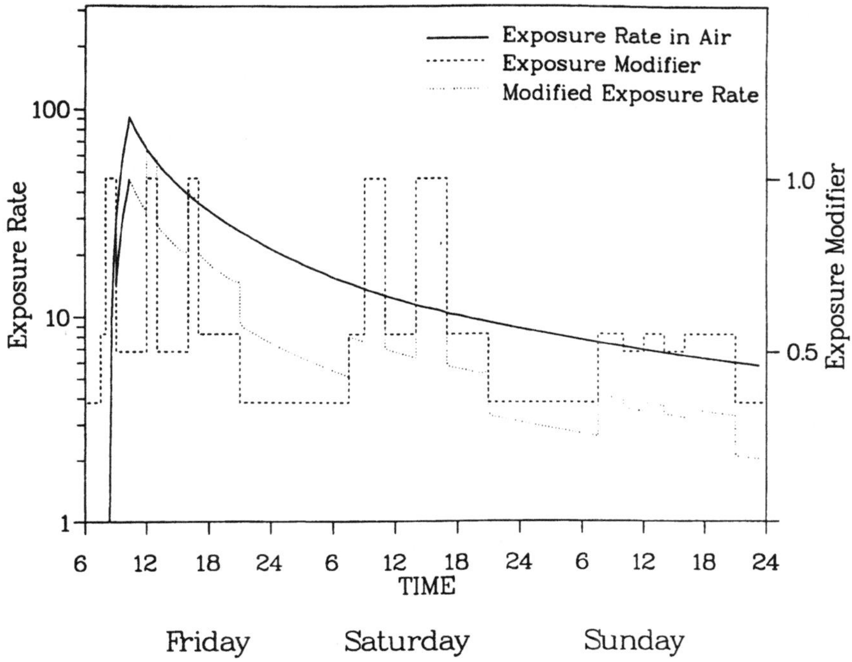


*Figure 4.* Composite curve presenting the free-in-air exposure rate, exposure modifiers used to represent shielding situations, and the modified exposure rate used in the γ-dose calculations.

The dotted curve in Fig. 4 is the actual function integrated. This detailed calculation is carried out for 30 d, with the result used to extrapolate to the present. This result is the shielded exposure for the individual from the time of cloud arrival until the present. The exposure is converted to whole-body dose in rads (for historical reasons) using a conversion factor derived from Ashton and Spiers (1979).

Beta skin doses were also calculated during the 30 d. If the individual was assumed to be outdoors during cloud passage, an approximation of the skin contamination was calculated. This was a function of the ground concentration, the fraction of cloud passage outdoors, and a skin contamination factor calculated from data presented by Kochendorfer and Ulberg (1967). These skin contamination factors were calculated as a function of month and time of day, and ranged from 0.58 to 4.6. The skin dose rate from this direct contamination was integrated from the end of cloud passage until it was assumed to be washed off by bathing.

During the period of direct exposure to the cloud, the β-particle dose to the skin from immersion was estimated. This was a small component of the total skin dose, since the direct contamination was happening at this same time, but was included for completeness.

For each period outdoors, the β skin dose from material deposited on the ground was calculated. This component represented a larger or smaller fraction of the skin dose, depending on the extent to which the individual was contaminated during cloud passage.

As shown in Figs. 1 and 4, the model requires an exposure modifier for the bedroom and living areas of the home. A data base was generated for these modifiers using representative data for homes extant during the period of atmospheric testing. During a tour of the southern part of Eastern Nevada and the southwest corner of Utah, 447 homes were visually examined and selected for inclusion in the data base if they appeared to have been there since about 1950. For each selected home, the approximate size, roof type, apparent construction, and assumed location of bedrooms were recorded. The location of the home with regard to neighbors was also recorded. Exposure modifiers were calculated for the bedroom and living areas of the homes using methods presented by Spencer (1962). A central value and a measure of its dispersion were calculated for the living area and bedroom of each home. These results were tabulated and used in that form as input for the model.

The basic input data set for the code was an ordered collection of the locations where fallout data were recorded by monitors in the field or for locations included within the bounds of a fallout map. This is the Town Data Base developed by the Desert Research Institute and reported by Thompson (this issue). For each such location, the data set included the name of each event that deposited significant fallout on the location, estimates of the cloud arrival time, and the normalized exposure rate in mR $h^{-1}$ at H + 12 h, along with expressions of the uncertainties in the estimates. Each record also included an estimate of the population, by life-style, at the time of the event.

For each event at each location, the code made 25 estimates of the whole-body $\gamma$ dose and $\beta$ skin dose for each of the nine life-styles. For each estimate, the stochastic portion of the code randomly selected a cloud arrival time, a normalized exposure rate, and a house from the list of 447. Using these data, the code stochastically followed the representative of the life-style from change to change for 30 d. During this period, estimates of $\gamma$ exposure, whole-body dose, and $\beta$ skin dose were made and accumulated. This period accounted for 75-90% of the total whole-body dose. depending on cloud arrival time. It was further assumed to account for 100% of the $\beta$ skin dose.

The result of the shielded and unshielded $\gamma$ exposures were used to derive an estimate of the whole-body dose for the time from the end of the 30 d to the present. The ratio of the two values was formed, the unshielded exposure calculated from the end of the 30 d to the present, and the ratio used to calculate the shielded exposure.

During the calculations for the homemaker life-style, an estimate of the uterine dose was also made. This estimate was used in combination with the exposure decay rate to produce estimates of uterine dose for the next 8 mo.

The 25 individual estimates of whole-body $\gamma$ dose, $\beta$ skin dose, and the monthly uterine dose were used to derive geometric mean and standard deviations. These 25 estimates were also written to a file for subsequent processing. The 25 estimates were used to accumulate information for multiple events at a location. Thus, the reported results (geometric means) for multiple events at a location are the geometric means of 25 sums of estimates, not sums of geometric means.

### 3. Results and Discussion

The model was run extensively for litigants in the Allen vs. U.S. trial. Since these calculations were limited to a few individuals at a few locations, 200 passes were made throughout the calculational parts of the code for each event. These results were subjected to a series of statistical analyses and found to more nearly approximate a lognormal than a normal distribution. These results of the analysis were essentially the same for all events, life-styles, and locations. Sequential analysis of the set of results indicated that little change in the geometric standard deviation was observed after about 20 passes. These findings led to the determination to use 25 passes through the code and to report all results as geometric means and standard deviations. (Note that all summing is done using the individual results; the geometric means of the sums are reported.)

The primary output of the program was a listing of individual dose estimates for all nine life-styles for each event at each location where significant fallout was detected. The first tabulation contained the results ordered by location. This included the population, name of event, and results. Estimates were given for the geometric mean of the whole-body $\gamma$ doses for each of the nine life-styles, along with the geometric standard deviation of the estimate. This same information was then presented for the $\beta$ skin dose. Next, the estimates for the uterine doses, as a function of month following the event, were given. These results were derived for all entries in the Town Data Base.

Figure 5 shows the results of dose calculations for Bunkerville, NV, following event SIMON, a 43-kt shot detonated on a tower. This shows the geometric mean and 2 standard deviations for doses for each of the nine life-styles. The figure shows, as expected, that the less-shielded life-styles are estimated to receive greater whole-body doses, but the dose is not a strong function of life-style.

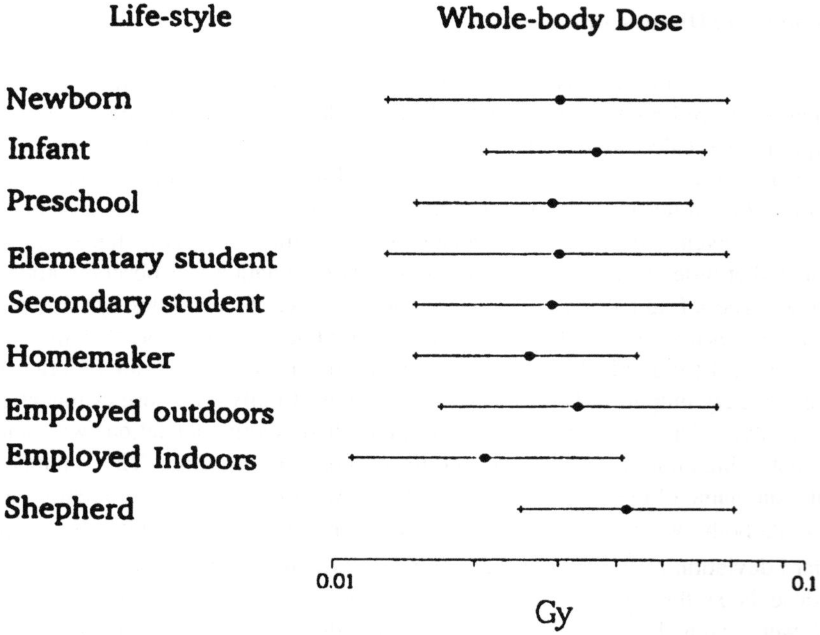


*Figure 5.* Estimated whole-body γ doses at Bunkerville, NV, by life-style following passage of the SIMON debris cloud. The central value is the geometric mean of the whole-body γ-dose distribution, and the error bars represent two geometric standard deviations (geometric mean and two standard deviations).

Tables 1 and 2 compare results produced by the external dose model with similar results presented by Anspaugh and Church (1985). Table 1 gives a comparison of the results for Bunkerville, NV, compiled by the Test Manager's Committee to Establish Fallout Doses (TMCEF) presented by Anspaugh and Church (1985) in Appendix C and modified results from the external dose assessment model. The original data are listed as "Estimated Radiation Doses [Roentgens]." The external dose model calculated exposures in roentgens as an intermediate step to estimating whole-body dose in rad. These results are given in the table to maintain a historical perspective. As noted above, the results labeled *All Events for the External Dose* are the geometric means of 25 independent summations over the eight events of those values.

TABLE 1. Comparison of results of TMCEFD and the external dose model at Bunkerville, NV.

| Event | TMCEFD | Individual exposures (R) External dose Employed indoors | Shepherd |
|---|---|---|---|
| ANNIE | | 0.0026 | 0.0080 |
| SIMON | 4.0 | 2.8 | 5.6 |
| HARRY | 0.20 | 0.050 | 0.12 |
| HORNET | 0.02 | 0.015 | 0.026 |
| ESS | 0.02 | 0.0047 | 0.0079 |
| ZUCCHINI | 0.06 | 0.075 | 0.16 |
| PRISCILLA | | 0.0068 | 0.012 |
| SMOKY | 0.17 | 0.12 | 0.17 |
| Hardtack II | 0.05 | | |
| All Events | 4.5 | 3.2 | 6.3 |

TABLE 2. Comparison of results of Beck and Krey, the TMCEFD, and the external dose assessment model.

| Utah location | Beck and Krey | TMCEFD | Estimated exposure (R) External dose Employed indoors | Shepherd |
|---|---|---|---|---|
| Beaver | <0.42 | 0.25 | 0.23 | 0.40 |
| Cedar City | 0.42 | 0.64 | 0.43 | 0.78 |
| Enterprise | 1.2 | 0.79 | 0.55 | 0.97 |
| Hurricane | 2.9 | 3.5 | 3.2 | 5.7 |
| Kanab | 0.49 | 1.6 | 1.2 | 2.6 |
| Kanarraville | 0.49 | 1.9 | 1.5 | 2.5 |
| La Verkin | 2.9 | 3.7 | 2.5 | 4.5 |
| Milford | <0.42 | 0.10 | 0.12 | 0.19 |
| Minersville | 0.69 | 0.20 | 0.26 | 0.44 |
| Modena | <0.42 | 0.54 | 0.41 | 0.74 |
| Panguitch | 0.28 | 0.70 | 0.47 | 0.80 |
| Paragona | 0.77 | 0.42 | 0.22 | 0.38 |
| Parowan | 0.77 | 0.42 | 0.29 | 0.46 |
| St. George | 2.6 | 3.7 | 3.0 | 5.8 |
| Santa Clara | 1.7 | 4.3 | 2.5 | 4.7 |
| Veyo | 4.1 | 2.8 | 2.0 | 3.9 |
| Washington | 1.7 | 3.3 | 2.8 | 5.0 |

Data in Table 2 show a comparison of the estimated exposure data generated by H. Beck and P. Krey and the Test Manager's Committee to Establish Fallout Doses as presented by Anspaugh and Church (1985) in Table 7, and modified results from the external dose assessment model. Results for the employed indoors and shepherd life-styles are shown to represent the range of results from the external dose code. Tables 1 and 2 demonstrate the general good agreement among these methods of exposure and dose estimation.

## 4. References

1. Anspaugh. L. R.; Church. B. W. Historical estimates of external gamma exposure and collective external gamma exposure from testing at the Nevada Test Site. I. Test series through HARDTACK II. 1958. Las Vegas, NV: U.S. Department of Energy, Nevada Operations Office; NVO-226; NVO-226;1985.
2. Ashton. T.; Spiers. F. W. Attenuation factors for certain tissues when the body is irradiated omnidirectionallv. Phys. Med. Biol. 24:950-963; 1979.
3. Hicks. H. G. Results of calculations of external gamma radiation exposure rates from fallout and the related radionuclide compositions. Livermore, CA: Lawrence Livermore National Laboratory; UCRL-53152. Parts 1 thru 8; 1981.
4. Hine. G. J.; Brownell. G. L. Radiation dosimetry. Third printing. New York: Academic Press. Inc.: 1956.
5. Kochendorfer. D. B.; Ulberg. .I. C. Human exposure to particulate debris from nuclear applications. San Francisco, CA: Naval Radiologicai Defense Laboratory; USNRDL-TR-67-59; 1967.
6. Kocher. D. C. Dose-rate conversion factors for external exposure to photon and electron radiation from radionuclides occurring in routine releases from nuclear fuel cycle facilities. Health Phys. 38:543-62 1; 1980.
7. Spencer. L. V. Structure shielding against fallout radiation from nuclear weapons. Washington, D.C.; National Bureau of Standards: NBS MONO 42; 1962.
8. Thompson. C. B. Estimates of exposure rates and fallout arrival times near the Nevada Test Site. Health Phys. 59:555-563; 1990.

# COMPARATIVE PLUTONIUM-239 DOSE ASSESSMENT FOR THREE DESERT SITES:
***Maralinga, Australia; Palomares, Spain; and the Nevada Test Site, USA—Before and After Remedial Action***

B.W. CHURCH
*BWC Enterprises, Inc.,*
*Logandale, NV, USA*

J.H. SHINN
*Lawrence Livermore National Laboratory,*
*Livermore, CA, USA*

G.A. WILLIAMS, L.J. MARTIN, R.S. O'BRIEN
*Australian Radiation Protection and Nuclear Safety Agency,*
*Yallambie Victoria, Australia*

S.R. ADAMS
*Bechtel Nevada,*
*Las Vegas, NV, USA*

## 1. Introduction

As a result of nuclear weapons testing and accidents, plutonium has been distributed into the environment. The areas close to the sites of these tests and accidental dispersions contain plutonium deposition of such a magnitude that health authorities and responsible officials have mandated that the contaminated areas be protected, generally through isolation or removal of the contaminated areas. In recent years remedial actions have taken place at all these sites. For reasons not entirely clear, the public perceives radiation exposure risk to be much greater than the evidence would suggest [1]. This perception seems to be particularly true for plutonium, which has often been "demonised" in various publications as the "most hazardous substance known to man" [2]. As the position statement adapted by the Health Physics Society explains, "Plutonium's demonisation is an example of how the public has been misled about radiation's environmental and health threats generally, and in cases like plutonium, how it has developed a warped 'risk perception' that does not reflect reality" [3].

As a result of this risk perception and ongoing debate surrounding environmental plutonium contamination, remedial action criteria are difficult to establish. By examining the data available before and after remedial actions taken at the three sites discussed in our report, we hope to present data that will illustrate that plutonium measured as aged deposition (older than several months) does not present as high a dose



*S.S. Hecker et al. (eds.), Nuclear Physical Methods in Radioecological Investigations of Nuclear Test Sites, 203-219.*
© 2000 *Kluwer Academic Publishers. Printed in the Netherlands.*

or risk as many had expected. The authors show that even though dose to the lung from inhalation (the primary pathway for the high-fired plutonium oxide particles present at these sites) is reduced, such a reduction is achieved at significant cost. The cost comes from damage to the environment, large expenditures per hectare rehabilitated, and the risk to occupational workers.

This paper specifically examines sites that are similar in many ways. These sites were chosen for their similarities to make comparisons. The sites are all desert in nature i.e., have low rainfall (all receive about 20 cm per year), have minimal vegetative ground cover, and have high summer temperatures. These sites are Palomares, Spain; the Nevada Test Site (NTS); and the Maralinga site in Australia. One significant difference, however, is that the Palomares site has been used continuously for residential and agriculture purposes since the plutonium remediation was completed. Maralinga is being remediated with the objective of returning the land to its former owners, but it will have some use restrictions for the remaining contaminated areas. Any decision to return the land being remediated by the United States Department of Energy (USDOE) at its Nevada sites, for public use, is in the distant future.

The plutonium source terms for each site are similar in that they were distributed as a result of the high explosive (HE) (i.e., chemical explosives) being detonated either purposefully or by accident. The resulting plume of vaporised plutonium was distributed in a cigar-shaped pattern dictated by the prevailing winds at the time of detonation. From this origin, the source material has the common characteristics of being created through a high-temperature process and rapidly oxidising to a very stable chemical state.

This high-fired Pu oxide is highly insoluble in the gastrointestinal system [4] and passes rapidly through the body when ingested. This accounts for the very small dose conversion factors for ingestion published by standard setting organisations such as the International Commission on Radiological Protection (ICRP) [5]. As a result of the nuclear nature of Pu-239, it decays primarily with a 5.15 MeV alpha emission, which means that a Pu-239 atom must be in close proximity to a cell before it can deliver any energy to that cell and potentially do harm.

This last finding is another reason why the ingestion pathway is of very minor concern. The epithelial cells lining the gastrointestinal (GI) system are replaced fairly rapidly. That fact coupled with the short residence time of the plutonium atoms in the GI system results in a dose that is very small and negligible. However, the situation is different in the lung, where small plutonium particles (e.g., $< 5\ \mu m$) can have very long residence times [4], and can also be in close contact with the lung's alveoli cells. This fact makes the lung the organ of concern when exposure to occupants of plutonium-contaminated land or re-occupants of Pu-remediated land are considered.

The route of exposure to the lung is through the inhalation process and an understanding of the factors and parameters that influence this pathway is necessary to predict the dose that an occupant of the contaminated land may receive. The primary parameters to be understood are source magnitude, particle size, and the combination of area size, wind speed, vegetative cover, and soil characteristics that give rise to the resuspension potential of the Pu-239 particles. Resuspension is largely an area phenomenon that requires a fairly large area affected by the contamination of interest, that is, an area where the contamination is on or very near the soil surface. Given these conditions when the wind blows, the particles can be picked up and consequently form an air stream of significant plutonium contamination. Several workers have discussed these dynamics and they have shown that receptors down wind will only be influenced by surface material within a certain distance. Shinn and Gouveia [35] in their report, "The Footprint Area Influencing a High Volume Air Sampler," have clearly demonstrated how the footprint fetch impacts a potential receptor. The worst case situation discussed in this report is bare soil, indicating that for a sampler at 1.13 meters high, 90% of the representative flux can be influenced by particles coming from as far away as 175 meters. When soil cover is present, the distances are reduced. Because of this area influence, measuring or determining Pu soil activity as an area average, i.e., $Bq/m^2$, is an important consideration.

The relationship of the above parameters is clearly discussed in many published reports, but we have chosen to follow the dose calculation method presented in the National Council on Radiological Protection and Measurements report (NCRP) 129 [6]. The model for lung inhalation dose used in that NCRP report is presented in Tables 1 and 2 along with the definition of the various parameters.

## 2. Methods

This paper focuses on a comparison of what has been learned from the measurements taken by various workers at these three desert sites to evaluate the resuspension parameters and how the remedial measures have affected them. We have examined the data available for each site and have compiled the data collected where available both before and after the remedial action to evaluate the effectiveness of the action. In the case of the Palomares site, which involved an accident, the data collected immediately after the contaminating event are very instructive.

Variables, which can be site and exposure scenario specific, such as the amount of air inhaled ($m^3$), along with the dose conversion factor (Sv/Bq), have been held constant for each of the before and after comparisons. We have used 8400 $m^3$ as the annual volume inhaled for adults living outdoors. We chose this number because the Maralinga Aborigines living a semitraditional life style live outdoors. The only change in the use of this parameter comes from calculating the dose for the lifestyle of hunting and hiking permitted to the returning Aboriginals. The volume inhaled has been adjusted to reflect the time individuals might spend in a worst case situation, which in this case would occur to individuals riding in the back of a following pickup truck while they were hunting kangaroo. As implied above in the case of the NTS and the Maralinga site, the calculated doses are hypothetical, as though people had been exposed or will be exposed at the levels indicated. However, this is not the case at Palomares. There, the values used and discussed have been experienced and continue to be experienced by the actual people who have been monitored (urine sampling and whole-body (i.e., lung) counting) for many years.

TABLE 1. Comparative dose assessment at the Maralinga and Palomares desert sites before and after soil remediation

| | A | B | C | D | E | F | G | H | I |
|---|---|---|---|---|---|---|---|---|---|
| 1 | PARAMETER | Maralinga, Au before cleanup (Central Taranaki) and | (Site FN) | Maralinga, Au after cleanup (Site FN) and | (West Plume hunting scenario w/restrictions) | Palomares, Sp before cleanup | Palomares, Spain after cleanup (loc 2-2, 15y ave) | (loc 2-1, 4y ave) (9,10,31) | (loc 2-1, 1993) |
| 2 | S--Activity in Soil (Bq/g) | 21 (24) | 4.2 (34) | 1.3 (25) | 0.74 (23) | 67 (10) | 2.06 | 0.44 | 0.44 |
| 3 | $E_r$-- (ratio between activity in respirable fraction of resuspended soil and activity in surface soil) | 20 | | | 6 | | | | 0.28 (31) |
| 4 | Arf -- Respirable fract. of resuspended soil- (Bq/g) **[D--soil deposition Bq/m$^2$]** | 420 | **[5.0 E+5] (34)** | **[2.0 E+4]** | 4.44 | **[1.2 E+6]** | | **[5.0 E+5]** | 0.123 **[3.3 E+4]** (31) |
| 5 | M--(mean dust loading in air (g/m$^3$) **[$S_f$-resuspension factor- m$^{-1}$]** | 2.0 E-5 (34) | 1.0 E-5 **[6.0 E-11]** | 2.2 E-5 **[4.7 E-9] (windy conditions)** | 0.08 | **[1.0 E-7t+init] [1.0 E-9t+2m] [1.0 E-10t+yrs]** | | **[2.6 E-10]** | 7.0 E-5 **[1.0 E-10]** (31) |
| 6 | $C_{air}$ -- Activity in Air (Bq/m$^3$) | 8.4 E-3 | 3.0 E-5 | 9.1 E-5 | 3.6 E-3 | 1.2 E-1 | 5.36 E-5 | 5.6 E-5- | 8.6 E-6 (31) |
| 7 | R -- (inhalation rate [m$^3$/y]) (26) | 8.4 E+3 | 8.4 E+3 | 8.4 E+3 | 672.0** | 8.4 E+3 | 8.4 E+3 | 8.4 E+3 | 8.4 E+3 |
| 8 | AIA--Annual Intake of activity (Bq/y) | 70.6 | 0.25 | 0.76 | 2.4 | 1008 init. 10.1 t+2m 1.0 t+yrs | 0.45 | 0.47 | 0.072 |
| 9 | $Df_{inh}$--[effective dose per unit intake (Sv/Bq)] | 5.7 E-5* | 5.7 E-5 | 5.7 E-5 | 5.7 E-5 | 5.7 E-5 | 5.7 E-5 | 5.7 E-5 | 5.7 E-5 |
| 10 | equals $E_{inh}$--[effective dose from 1 yr of intake] (S v y$^{-1}$) | 4.0 E-3 | 0.014 E-3 | 0.043 E-3 | 0.14 E-3 | 57.5 E-3 init. 0.57 E-3 t+2m 0.057E-3 t+yrs | 0.026 E-3 | 0.028 E-3 | 0.0041 E-3 |
| 11 | $E_{inh}$ in mSv/y | 4 | 0.014 | 0.043 | 0.14 | 57.5 init. 0.57 t+2m 0.057 t+yrs | 0.026 | 0.028 | 0.004 |
| 12 | Dose Model--$E_{inh} = Df_{inh} \times C_{air} \times R$; $C_{air}$ can be measured directly or estimated by $C_{air} = E_r \times S \times M$ or by the Resuspension Factor method where $C_{air} = S_f \times D$ (6)<br>* The dose conversion factor ($Df_{inh}$) is based on NRPB experimental work on Maralinga Pu particals; NRPB recommended that 75% Y & 25 % W of the ICRP values be used (26). This value of 5.7 E-5 Sv/Bq has been used for all calculations for comparison reasons.<br>** Exposure while hunting (27) | | | | | | | | |

TABLE 2. Comparative dose assessment at the Nevada Test Site before and after remediation

| | A | B | C | D | E | F |
|---|---|---|---|---|---|---|
| 1 | PARAMETER | Nevada Test Site, USA<br>Double Tracks<br>before and | after cleanup | Nevada Test Site, USA<br>Clean Slate I<br>before and | after cleanup | Reference |
| 2 | S--Activity in Soil (Bq/g) | 4.52 Wtd. Ave. (1) | 1.40 Wtd. Ave. (1) | 4.81 Wtd. Ave. (1) | 0.57 Wtd. Ave. (1) | (1) Steve Riedhauser, BN 29 Apr 99 (ref. 36) |
| 3 | $E_r$-- (ratio between activity in respirable fraction of resuspended soil and activity in surface soil) | 0.69 (3) | 0.69 (3) | 0.69 (3) | 0.69 (3) | (2) Desert Research Institute Air Sampling Report for Clean Slate I, 1998 (ref. 37) |
| 4 | Respirable fract. of resuspended soil-(Bq/g)<br>**[D--soil deposition Bq/m$^2$]** | **[6.8 E+4 Wtd. Ave. (1)]** | **[2.1 E+4 Wtd. Ave. (1)]** | **[7.2 E+4 Wtd. Ave. (1)]** | **[8.6 E+3 Wtd. Ave. (1)]** | (3) Shinn-1997 Resuspension at the Tonapah Range (ref. 29) |
| 5 | **M**--(mean dust loading in air (g/m$^3$)<br>**[$S_f$ resuspension factor- m$^{-1}$]** | 5.9 E-6 (4)<br>**[1.3 E-10 (3)]** | 8.2 E-6 during (4)<br>4.8 E-6 after (4)<br>**[1.3 E-10 (3)]** | 3.5 E-6 (2)<br>**[1.3 E-10 (3)]** | 7.1 E-6<br>**[1.3 E-10 (3)]** | (4) Desert Research Institute Air Sampling Report for Double Tracks. 1997 (ref. 33) |
| 6 | **$C_{air}$** -- Activity in Air (Bq/m$^3$) | 1.1 E-7 (5) measured<br>**[8,8 E-6 calc. from $S_f$ & D]** | 2.4 E-5 meas during (4)<br>1.11 E-7 meas. after (5) [2.7 E-6 calc fm $S_f$ & D] | 7.2 E-6 meas. (2)<br>[9.4 E-6 calc fm S & D] | 1.6 E-4 meas. during (2)<br>2.8 E-6 meas. after (2)<br>**[1.11 E-6 calc after]** | (5) DOE/NV Double Tracks Closure Report-1997 (ref. 32) |
| 7 | R -- (inhalation rate [m$^3$/y]) (26) | 8400 | 8400 | 8400 | 8400 | Inhalation rate for standard man living outdoors 365 days per year (26) |
| 8 | AIA--Annual Intake of activity (Bq/y) | 9.2 E-4 fm meas.<br>**[7.4 E-2 fm $S_f$& D]** | 2.0 E-1 during (4)<br>9.3 E-4 after (5)<br>[2.3 E-2 calc fm S & D] | 4.03 E-2<br>[7.9 E-2 calc. fmS & D] | 1.34 during excav. (2)<br>2.3 E-2 after excav. (2)<br>[9.3 E-3 calc after, S& D] | Note: At Double Tracks and Clean Slate I, DRI collected air samples and dust samples at the edges of the area undergoing excavation in the direction of the prevailing winds. Sampling was done months before, during and after the excavation period. Water was used to supress dust during excavation and a fixative was applaid after. These methods/agents were fairly effective as the observed $E_{inh}$ is very small. The during values being the only ones showing any modest resuspension. |
| 9 | **$Df_{inh}$**--[effective dose per unit intake (Sv/Bq)] | 5.75 E-5 * | 5.75 E-5 * | 5.75 E-5 * | 5.75 E-5 * | |
| 10 | **$E_{inh}$**--[effective dose from 1 yr of intake] (S v y$^{-1}$) | 5.2 E-8 calc fm meas.<br>**[4.2 E-6 calc.]** | 1.1 E-5 during excav.<br>5.3 E-8 after excav.<br>[1.3 E-6 calc fm S & D] | 2.3 E-6 before<br>[4.5 E-6 before, S & D] | 7.6 E-5 dur excav<br>1.3 E-6 after excav<br>[5.3 E-7 after, S&D] | |
| 11 | **$E_{inh}$ in mSv/y** | 5.2 E-5<br>[4.2 E-3 calc.] | 1.1 E-2 during excav.<br>5.3 E-5 after excav.<br>[1.3 E-3 calc fm S & D] | 2.3 E-3 before<br>[4.5 E-3 before, S & D] | 7.6 E-2 dur excav<br>1.3 E-3 after excav<br>[5.3 E-4 after, S&D] | |
| 12 | Dose Model--$E_{inh} = Df_{inh} \times C_{air} \times R$; $C_{air}$ can be measured directly or estimated by $C_{air} = E_r \times S \times M$ or by the Resuspension Factor method where $C_{air} = S_f \times D$ (6)<br>* The dose conversion factor ($Df_{inh}$) is based on NRPB experimental work on Maralinga Pu particals; NRPB recommended that 75% Y & 25 % W of the ICRP values be used (26).<br>This value of 5.7 E-5 Sv/Bq has been used for all calculations for comparison reasons. | | | | | |

## 3. The Three Sites

### 3.1. PALOMARES, SPAIN

On January 17, 1966, a U.S. Air Force B-52 bomber collided with its tanker and exploded above the town of Palomares, Spain. Of the four nuclear weapons onboard the bomber, three impacted very near the town and the fourth fell into the sea. The chemical high explosives of two of the four weapons detonated on impact and bracketed the residential area. Plutonium particulate contamination was distributed in varying degrees over a 226 ha area (~560 acres) consisting of brush land, farmland, and an urban area. The other two weapons were recovered intact.

A set of tiered criteria applied to the cleanup (Table 3) was completed by May 1966. In the most heavily contaminated area, crops and soil were removed to a depth of approximately 10 cm. These materials were then packaged and shipped to the United States for disposal. The next level of contamination required that canes be burned on the beach, crops buried, and soil plowed to 30 cm. The remainder of the contamination required that soil be plowed to 30 cm where possible [7].

TABLE 3. Cleanup criteria for the three desert sites

| Site | Principal Criterion | Comments |
|---|---|---|
| Palomares, Spain<br>1/17/66; //66 *<br>(7) | 1180 $kBq/m^2$ -soil removed to 10 cm<br>118 $kBq/m^2$ -soil plowed to 30 cm<br>11.8 $kBq/m^2$ -some soil plowed to 30 cm | Approximately 1000 $m^3$ of soil was placed in 4810 barrels and shipped to the USA. Crops were removed and burned or buried on the beach<br>226 ha remediated |
| Maralinga, Australia<br>(British Test Site)<br>9/56-4/63; //67//96-99<br>(7) | part./frag.--remove if >100 Kbq;<br>if ≥ 20kBq averaged<br>Not More Than 1/10 $m^2$;<br>TRU in Soil (1ha ave.) / $Am\text{-}241/m^2$<br>Taranaki - 1.2 Bq/g / 3 $kBq/m^2$<br>TM-100 - 2.2 Bq/g / 1.8 $kBq/m^2$<br>TM-101 - 1.7 Bq/g / 4.0 $kBq/m^2$<br>Wewak - 0.7 Bq/g / 1.8 $kBq/m^2$ | activity of the part./frag. are for Am-241;<br>TRU is primarily Pu-239,240; The guideline varies primarily because of the difference in the Pu/Am ratio at the specific sites.<br>147.4 ha remediated |
| Nevada Test Site<br>Double Tracks Event<br>5/63; // summer 96<br>(17) | DOE Order 5400.5, Ch.4<br>1 mSv/yr (100 mrem/yr)<br>Remove Soil >7.4 Bq/g (200 pCi/g)<br>Pu-239,240; ave. over 1ha | Remove all fragments<br>2.6 ha remediated |

*Operational period or event date; **//69 Year or period of time cleanup took place. Some locations had more than one cleanup.

Upon completion of the cleanup and soil restoration of the farmland, a research/monitoring programme began. For reference, the contaminated area has been subdivided into three zones (zones 2, 3, and 5), corresponding to the number assigned to the weapons and where they impacted. Zone 5 is the urban zone (Fig. 1). This programme consisted of air sampling, soil sampling, crop sampling, urine sampling, and lung counting of residents. Members of the Spanish Junta De Energia Nuclear [8, 9, 10] carried out this work.

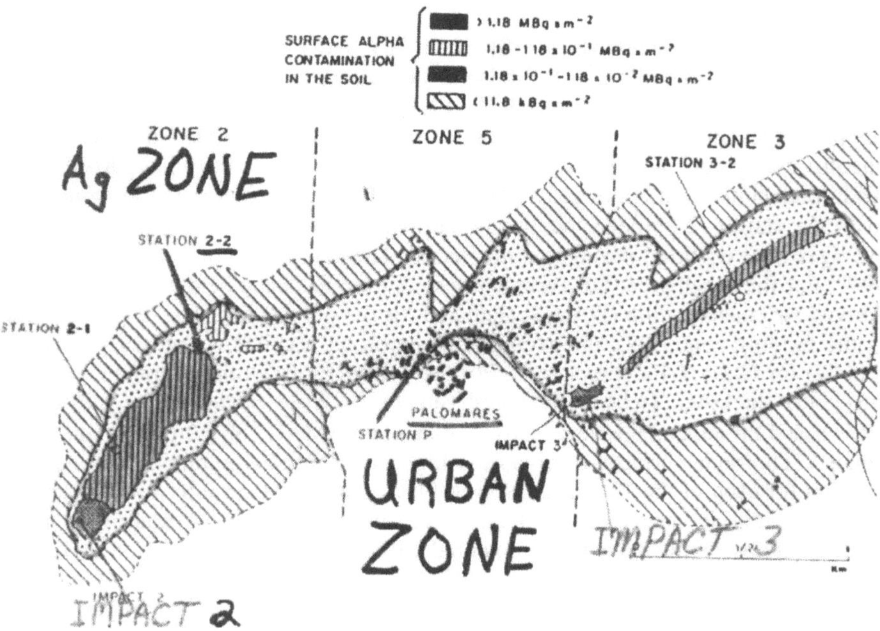


*Figure 1*. Palomares area with original surface contamination levels and location of air sampling stations. Black spots show the location of houses.

On the day of the accident 485 people were present in Palomares. However, because it was a holiday, few were working in the nearby fields. Over the 22 years of monitoring by the Spanish Government, 229 additional people have been included, bringing the total to 714. They either moved into the area or were born after the accident. Of this number 590 had urine sample results lower than the minimum detectable activity (MDA) of 0.37 mBq/d and 124 had values equal to or higher than the MDA. Of the 124, 29 showed sample contamination, because samples were collected in Palomares. Beginning in 1967, residents travelled to Madrid for sampling and annual examination. Estimations of dose were performed on those 55 people who were considered to really have suffered internal contamination. Lung counting, using the most sophisticated detectors for the time, indicated no plutonium level above the MDA of 814 Bq. From the urine sampling, it was determined that 45 of the people had experienced acute inhalation of particles. Chronic inhalation at the average concentration measured in air could not support the quantity observed in the urine. From this the date of intake could be assigned. For the remaining 10, who were not in the area on the day of the accident, the date of intake considered to be the most likely was assigned. The 70-year whole-body dose committed effective dose equivalent (CEDE) for the initial 45 residents ranged from 20 to 200 mSv (0.29-2.9mSv/y). On the basis of the suppositions for the group of 10, a range of from 35 to 180 mSv is estimated. The remaining population

(659 residents) is estimated to have received less than 20 mSv, CEDE. There were 10 residents who were 15 years or younger. Of these only one (<1 yr on Jan. 17, 1966) has a CEDE higher than 200 mSv (242 mSv); the other 9 are in the range of from 49 to 157 mSv [11].

Aerosol measurements [8, 9, 10] of Pu-239 and Pu-240 (Table 4 and Figure 2) exceeded the detection limit (1.8 E-6 $Bq/m^3$) for all years at all sampling stations, except in the urban zone for the years 1971, 1972, 1975, and 1976. The frequency of air samples exceeding the detection limit diminished with time until a hilly and uncultivated parcel was plowed and transformed into a cultivable parcel in 1974. Additional cultivation in the early 1980s also contributed to increasing air concentration [31]. At station 2-2, near the newly cultivated land, higher annual concentration averages have occurred throughout the period. The maximum concentration occurred in 1967. For the urban area, the average concentration has been below a hundredth of the derived air concentration (DAC) (calculated for the public) for Class Y Pu compounds. In the farming zone (2-2), the average concentration was below one-tenth of the DAC (public) for Class Y Pu compounds. For the period 1966-69, the average concentration was below the DAC. To be on the conservative side, doses calculated for various organs were based on an average mass activity diameter (AMAD) of 1.0 μm for the particles inhaled. The bone surfaces received the highest potential committed dose equivalent; the sum for each of the 15 years (1966-80) has a value of 0.56 mSv for the urban zone and 5.42 mSv in zone 2-2. In the five organs of interest, the contribution of the committed dose equivalents to the potential CEDE during the 15-y period is 0.054 mSv in the urban area and 0.52 mSv in zone 2-2 [9].

The Palomares region of Spain is typical of Mediterranean agricultural areas that receive marginal rainfall (~20 cm/y) and require irrigation to sustain crops. Common crops of the area are tomatoes, grain, and alfalfa. The plutonium concentration observed for washed tomatoes is 0.15 Bq/kg and is generally a factor of 30 to 40 higher for the plants, stalks, etc. of tomato plants, barley, and alfalfa. The soil-crop concentration ratios are of the order of $10^{-4}$ for tomatoes and $10^{-3}$ for the plants and the components of barley and alfalfa. The annual CEDE to individuals ingesting tomatoes is 1.5 μSv. Other pathways, i.e., alfalfa, meat, or milk, to humans are even less [12].

The data and experiences recorded for Palomares are very important. To date the people of Palomares belong to a very well-documented group, that since 1966 have lived and worked continually as farmers in a plutonium-contaminated environment.

TABLE 4. Composite Data Table, Palomares, Spain 1966-88

| Pu-239 + 240 in Samples or Dose from Pu 239 + 240 in Organs* | | | | | | |
|---|---|---|---|---|---|---|
| Parameter Reported | Urban Zone | | 2-2 Zone (Agr.) | | 2-1 Zone (Agr.) | |
| Soil surface contam. MBq/m2 | 0.012 | | 0.012 to 1.2 | | 0.012 to 1.2 | |
| Concentration in Soil (Bq/g Ave. Max. Min. (28) | 0.13<br>0.30<br>0.02 | | 2.06<br>3.31<br>0.80 | | 0.44<br>1.60<br>0.03 | |
| Sample Year | Surface Soil Conc. MBq/m$^2$ | Airborne µBq/m$^3$ | Surface Soil Conc. MBq/m$^2$ | Airborne µBq/m$^3$ | Surface Soil Conc. MBq/m$^2$ | Airborne µBq/m$^3$ |
| 1966 | 14.8 | | 44.8 | | | 41.8 |
| 67 | 4.1 | | 441.8 | | | 15.2 |
| 68 | 2.6 | | 21.8 | | | 7.0 |
| 69 | 0.0009 | 2.6 | 0.19 | 142.0 | | 161.0 |
| 70 | 5.9 | | 2.2 | | | |
| 71 | < 1.8 | | 2.2 | | 4 year average = 56.0 µBq/m$^3$ | |
| 72 | 0.013 | < 1.8 | 0.27 | 10.4 | | |
| 73 | 0.0004 | 2.2 | 0.01 | 3.0 | | |
| 74 | 4.1 | | 8.1 | | | |
| 75 | 0.0006 | < 1.8 | 0.06 | 16.3 | | |
| 76 | 0.002 | < 1.8 | 0.01 | 4.4 | | |
| 77 | 0.001 | 5.6 | 0.01 | 11.8 | | |
| 78 | 2.2 | | 16.7 | | | |
| 79 | 0.003 | 5.6 | 0.02 | 19.2 | | |
| 80 | 28.1 | | 32.9 | | | |
| 81 | 0.004 | 14.3 | 0.30 | 46.6 | 0.033 | 8.5 |
| (27,28) 83 | 0.004 | 9.1 | 0.31 | 87.9 | | |
| Shinn** * 1993 | | | | | | |
| Potential CEDE for 1st 15 years from chronic inhalation (1966-80) Five Organ Total** (mSv) | (lung-0.023)<br>0.054 mSv | | (lung-0.22)<br>0.522 mSv | | (lung-0.004 mSv/y Shinn, 1993) | |
| Percent of 1 mSv/y Effective Dose Equivalent (26,27) | 5.4 % | | 52.3 % | | | |
| (SE70 ) 70 y CEDE from the Acute Inhalation Exposure at the time of the accident for 45 residents (29) | 20-200 mSv (This data doesn=t apply to zones) | | | | ***Data from Shinn who independently completed a sampling trip in Zone 2-1 during 1993 | |

*Data from Iranzo, 1987 & 1994

Soil Surface Samples were collected using a square metal grid 25 cm x 25 cm x 5 cm in depth.

Soil samples were collected from study plots 50m X 50m designated for repeat sampling.

Aerosol sampling using high-volume samplers (1 m$^3$/min) were at various locations.

** Five Organs consist of: Lung, Bone Surface, Liver, Red Marrow, & gonads.

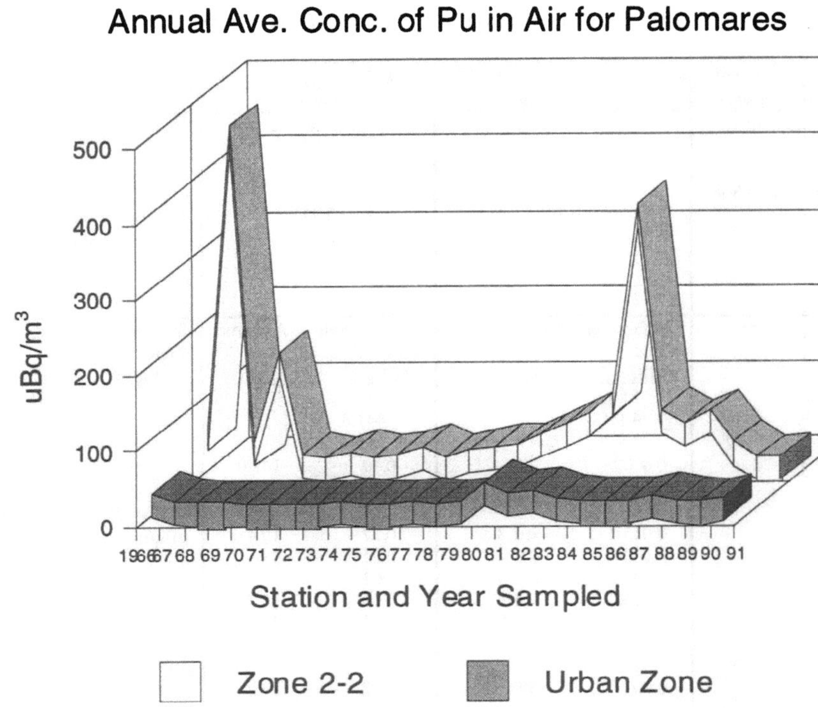


*Figure 2.*

## 3.2. NEVADA TEST SITE

The Nevada Test Site and environs have approximately 8500 acres contaminated with plutonium >3.7 Bq/g on the land surface because of one-point-safety tests and plutonium-dispersion tests [13]. During the summer of 1996, an interim cleanup action was completed at the Double Tracks event site. An additional site (Clean Slate 1) was cleaned up in the spring of 1997. Both were part of the Operation Roller Coaster series. These two sites conformed to the criteria outlined in Table 2. However, cleanup activities are considered by officials to be interim actions because no definitive plutonium-in-soil standard or guide exists. USDOE Nevada (NV) has recently negotiated a Federal Facility Agreement and Compliance Order (FFACO) with the State of Nevada. There has not been time for the State (lacking a federal plutonium-in-soil standard) to decide on a Nevada standard for USDOE to use. It has been contemplated, as suggested by the Double Tracks Environmental Assessment [14], that the cleanup on criterion, expressed as a concentration, is likely to be 14.8 Bq/g, but could be as low as 3.7 Bq/g. The value of 7.4 Bq/g as used is expected to be conservative [15].

The primary dose limit specified in USDOE Order 5400.5, Chapter 4, for all USDOE activities (including remedial actions), is expressed as a CEDE. This specification comes from the ICRP 26, 30 risk-based system, which requires a summation of doses to various organs of the body using weighting factors to be applied to each major tissue and organ. Exposure (as a consequence of USDOE activities) to members of the public from all radiation sources is not to exceed 1mSv/y. Though USDOE and others have established guidelines for thorium and radium in soil, guidelines for residual concentrations for other radionuclides in soil have to be derived from the basic dose limit. Creating an environmental pathway analysis, in which site data and default parameters are fed into a dose prediction model, does this. In USDOE's case they are required to use the RESRAD computer model developed by Argonne National Laboratory [16].

Because the "correct" land use and resultant pathway selection can be in the "eye-of-the beholder," this method generally causes considerable discussion among regulators, stakeholders, and management. This is happening with the Nevada State Division of Environmental Protection (NSDEP). The USDOE Nevada office had chosen to use a rancher's land-use scenario and resultant pathway assumptions [17], but the director of NSDEP has made it clear that he is not readily accepting the USDOE/NV land-use scenario [18]. He has implied that future negotiations must be held before the matter of the "correct" land-use and resultant cleanup criteria can be established.

### 3.3. MARALINGA, AUSTRALIA

Maralinga was one of three sites used by the British for nuclear weapons testing in Australia [19]. The Royal Commission Hearings in the mid-1980s initiated the events leading to the current remedial programme. A Technical Advisory Group (TAG) was formed upon a recommendation at the hearings to investigate the aspects of future remedial action and to recommend cleanup options for the Australian government to consider. The TAG report offered a group of options. From these, the government selected option 6c, which in brief, suggested surface-soil excavation and burial in deep trenches on site. Pits located a distance from Taranaki have been exhumed. Of the 21 shallow burial pits filled with plutonium-contaminated metal debris, 11 were treated by *in situ* vitrification; 10 were exhumed and their contents deeply buried. The posting of signs to limit activities in the large, unremediated downwind fallout plumes has been accomplished [20].

This project will cost about $A 104 million and will remove over 322 000 $m^3$ of soil [21]. The project's objective is to return use of government-controlled lands to former native owners. The soil-removal criterion for rehabilitation was based on the annual risk of fatal cancer associated with the inhabitants inhalation or ingestion of contaminated soil. This risk is not to exceed one in ten thousand accumulated by the fiftieth year. It was considered that the soil contamination contour corresponding to an annual committed dose of 5.0 mSv is the borderline between acceptable and unacceptable risk [20].

Large amounts of metal and other materials were immediately adjacent to the explosive device of each of the 12 one-point safe trials conducted at Taranaki. Because of these materials, an area of about 2.5 $km^2$ was extensively contaminated with fragments and particles (the difference being visible or not visible to the naked eye). Concern that the fragments might cause or enter a wound on the bare feet of the inhabitants largely dictated the extent of the area for soil removal. The remediation plan not only specified the various earth works, but it also had to deal with restrictions on land use. For example, in the plume areas where excavating such large areas was not feasible, only casual travel and hunting will be allowed. The option for the Aboriginal people to live in these areas was relinquished because of the cost and extensive environmental damage required to reduce the average plutonium concentration to an acceptable risk level for a semitraditional Aboriginal lifestyle. The Aboriginal landowners, who were able to participate in the development of cleanup objectives and plans, readily agreed to the hunting and travel restrictions. Their participation helped to prevent severe and extensive damage to the environment.

## 4. Results

Table 3 and Figure 2 display the monitoring data gathered for the Palomares site by Iranzo *et al.* [9,10]. These data are particularly compelling because, as has been mentioned, they are for a population that was not only present at the time of the incident, but one that has continued to live (now 33 plus years) in the contaminated area since the event. As was discussed above, only 45 individuals of those who were out of doors in the near vicinity of the HE explosive dispersal have had "measurable" plutonium in their urine. The remaining 440 have never had sufficient quantities in their excreted urine to detect plutonium. Over the years, these individuals, along with an additional 229 plus the miscellaneous exposed 10, add to 714 people who have lived and worked in and around the contaminated area illustrated in Fig. 1. The monitoring data displayed show low levels of airborne contamination with some years spiking because of increased disturbance of the nearby soil because of pond construction and the cultivation of new ground. However, even with these activities, the airborne concentration of plutonium remains low and the calculated dose, calculated because it is so low it cannot be directly measured via bioassay, is only a fraction of the dose published as acceptable and safe by international organisations [5, 22]. It is instructive to examine the data in Table 1, row 5, column F, which displays the resuspension factors measured by Iranzo *et al.* [10] at the time of the fresh deposition and then as it aged at 2 months and years. It is also observed in cell F-11 that the corresponding $E_{inh}$ (effective dose for inhalation) also drops dramatically.

At Maralinga the remediation, which is nearly complete, was driven by criteria developed for a returning Aboriginal community that would be living a semi-traditional lifestyle. These people live in a camping environment that creates considerable dust. The few dust loadings that were measured were high. The dust loading used in the dose-projection calculations was based on limited measurements indicating a value of 1.0 mg/m$^3$ for adults. A dust loading this large indicates a very dusty environment and is generally associated with major dust storms or severe dust-raising activities. The authors of NCRP report 129 present data from various workers who found annual averages of 28 and 75 $\mu$g/m$^3$ for undisturbed conditions for rural and urban locations, respectively. The report also shows data collected from behind a tractor, which had a median of 15 mg/m$^3$ and a range of from 0.3 to 200 mg/m$^3$ (mass loading values over 0.15 mg/m$^3$ are considered nuisance dust) [6]. The annual effective dose $E_{inh}$ calculations for the semitraditional Aboriginal lifestyle used 1 mg/m$^3$ dust-loading and enhancement factors as indicated in Table 1. The enhancement factor ($E_f$) is defined as the ratio of the activity concentration of the respirable fraction to the activity concentration of the sample as the whole. The $E_f$ was determined for soil samples from separations performed with a Bahco microparticle classifier. The $E_f$ for the resuspended material was determined from cascade impactor data. It is important to note that the "enhancement factor" as defined earlier contains no depth-related information. It merely relates the activity concentration in the inhalable fraction of a given soil sample to the activity concentration of the bulk sample, whether the sample is of surface soil or a suspension in air.

To convert from activity per unit surface area to activity per unit mass, it is necessary to know the profile of activity with depth, as well as the soil density. At Maralinga, measurements show that in undisturbed areas, most of the activity is in the top 10 mm of most samples and at least 85% is in the top 20 mm. Average soil density was found to be 1.7 g/cm$^3$. From these data and for the purposes of calculation, it is assumed that, in undisturbed areas, all the activity is in the top 10 mm of soil and that it is uniformly mixed throughout that depth. On the basis of these assumptions, an airborne dust concentration of 1 mg/m$^3$ corresponds to a resuspension factor of 6 x 10$^{-8}$ per m for an enhancement factor of unity.

Using the model parameters presented in Table 1, Williams *et al.* [23] calculated 273 mSv/y for central Taranaki using parameters specific to a camping lifestyle (i.e., large mass loadings). Johnson *et al.* [24] have published a dose commitment rate indicated in Table 1, in cell B-11 of 4 mSv/y for adults for central Taranaki. The difference can be attributed to undisturbed soil conditions at the time of the measurements as indicated by the low mass loading, 0.2 μg/m$^3$ (Table 1; cell B-5). The resuspension factor ($S_f$ ) determined by Johnson for a nearby area (site FN- Fig. 3) within central Taranaki before cleanup (cell C-5) is also very low. Soon after cleanup of this area, approximately 10 years later, a similar determination of $S_f$ was made by Shinn (cell D-5) [25]. The value in Table 1 is for windy conditions; the value for normal conditions is an order of magnitude lower (3.2 E-10 m$^{-1}$). Table 1, (cell D-5) also indicates a mass loading in the μg/m$^3$ range under windy conditions and very soon after excavation of the area was completed (i.e., under disturbed conditions).

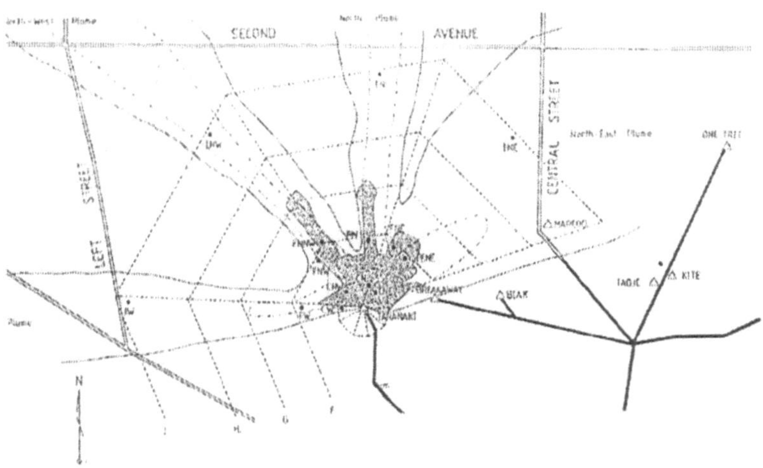


*Figure 3.* Map of Taranaki Area showing location of FN

Measurements and dose estimates have also been made in the downwind plumes using the model and aboriginal life style discussed above. The results of the calculation have been published by Heywood *et al.*, [26]. The input parameters were similar to those of Williams [23], e.g., mass loadings of 1 $mg/m^3$ for the adult and large enhancement factors, i.e., greater than unity. The value calculated is 3.0 mSv/y for the adult in the northwest plume; for the 10-year-old child it is about 20% more. In Table 1, column E, we have shown the results, using the same method as previously described, for the "hunting scenario."

The Aboriginal stakeholders have agreed to restrictions that limit their use of the land to walking and hunting, but no camping. The occupancy factor for this scenario has been adjusted according to the time spent hunting as determined by Palmer and Brady [27].

The USA has conducted tests at Nevada that were similar to the Vixen trials at Maralinga, and the USDOE has recently engaged in remedial actions at two of these event sites, Double Tracks and Clean Slate I of the Roller Coaster Series. The series consisted of four events, jointly sponsored by the US and UK, at the Tonapah Test Range (TTR), near Tonapah, NV. These tests were conducted to study the safety aspects of the transportation and storage of nuclear weapons. The data displayed in Table 2, attempt to show the changes in resuspension and the resulting $E_{inh}$ before and after the remedial actions. Like Maralinga, and unlike Palomares, the sites have not been used for many years. Prior to remediation the surface plutonium particles are in a stable and undisturbed desert shrub environment. As the data suggests, the various parameters indicating the ability to resuspend, i.e. $S_f$ and M were very low prior to the commencement of any earth works. Because the USDOE used water spray periodically to suppress dust during the excavation [32] and a soil fixative upon completion, the mass loading, M and $S_f$ remained low. Observable plutonium was measured on air samples placed at the perimeter on three sides covering the majority of the wind regimes recorded. However, the $E_{inh}$ would not exceed 10% of the annual effective dose to the public, if they stood at the excavation boundary continually for a year. The scenario is impossible because the excavation and shipping only lasted for less than a two-month period, after which the indicated $E_{inh}$ dropped by a factor of 58.

Resuspension studies conducted year around by Shinn *et al.*, [29] at TTR show that weathering, e.g., the spring freeze/thaw cycle, etc., does create variability in the $S_f$. The range is over an order of magnitude with the median at 1.3 E-10 $m^{-1}$. All the $S_f$ values measured at the three sites reviewed in this paper mostly fall within this range (1.0 E-9 to 1.0 E-10 $m^{-1}$). The exceptions were the initial observed $S_f$ at Palomares (E-7 $m^{-1}$) and the $S_f$ measured at central Taranaki before cleanup (E-11 $m^{-1}$). Shinn and others [30, 6] have also reported that this range is generally observed at many other sites around the world, e.g., at the Marshall Islands, at Chernobyl, and at other locations on the Nevada Test Site.

## 5. Conclusions

- Investigations of all three sites studied in this report show that within a short time period, e.g., a few months, plutonium-contaminated debris becomes fairly stable, even when the soil surface is periodically disturbed, as in the case of farming at Palomares, and of excavating at Maralinga and TTR. This finding is important in considering cleanup criteria, because it suggests that even after severe disturbances associated with remedial actions and or farming, the soil surface soon returns to a condition that limits resuspension of plutonium particles.

- From the experience and data observed at the three sites, it is possible to estimate conservatively the deposition required to deliver 1 mSv/y (annual dose commitment to a member of the public) to potential inhabitants who spend their entire year outside. The following are the required conditions: The $Df_{inh}$ is for very insoluble particles, such as high-fired Pu oxide; the $S_f$ falls within the range observed for plutonium material over several months old; and the mass loading is not something very abnormal ($\geq 1$ mg/m$^3$). With these general conditions met, one should be able to dependably solve for D (Soil Deposition, Bq/m$^2$). The solution indicates that it would require the soil deposition in the low MBq/m$^2$ range to come close to the dose limit for a member of the public.
- As Palomares is the only site where members of the public routinely live and work in a plutonium-contaminated environment, we learn that indeed working and living in these environments yield only fractions of the annual public effective dose. The data that Shinn [25] collected in 1993 in an agriculture area indicate less than 1% of the annual dose to a member of the public, and this is if they were at that location 365 days a year. To date it has not been reported that members of the community inhale or ingest enough plutonium from the contaminated soil to be detected in bioassay samples.
- If the Palomares experience is valid, and the $S_f$ levels observed around the world are reasonable indicators of potential exposure, then remedial actions looking at only a few Bq/g are very, very conservative (see Table 1).
- On the basis of the actual human experience observed at Palomares, Iranzo *et al.* [12] recommended that the International Atomic Energy Agency (IAEA) consider 1 man Sv of collective effective dose equivalent as a guideline to exempt quantities for practical application. They further recommend that for the crop types experienced, 120 to 1200 kBq/m$^2$ would be an appropriate intervention level, depending on the size of the contaminated area. The larger the area, the higher the intervention level.
- It is also observed for Palomares that only those individuals who were outdoors and in the near down wind area at the time of the accident received doses approaching and exceeding the annual dose limit for the public and had measurable plutonium in their urine samples. Resuspension of the deposited Pu particulate gives minimal dose to the receptor's organs. The CEDE is much less than the specified limits. Working, living, and eating products grown in a contaminated area of the magnitude experienced at Palomares yield extremely small doses that are much less than the accepted standards.

## 6. Acknowledgements

This to acknowledge and thank the Department of Industry, Science, and Resources of the government of Australia who supported the principal author and other work done in

Australia. Specifically, thanks to Mr. Jeff Harris, Dr. Caroline Perkins, and Ms. Sharon Burnell for their support. In addition, thanks to the USDOE, specifically Mr. Steve Mellington and Ms. Monica Sanchez for support of the data collection and sample analysis of Maralinga samples and for assistance in acquiring recent data and information about the Nevada sites.

## 7. References


1. J.D. Boice Jr., Risk Estimates for Radiation Exposures; In: Health effects of exposure to low-level ionizing radiation (Hendee W.R., Edwards F.M. eds. Philadelphia: Institute of Physics Publishing, 1996: 237-268.
2. Health Physics Society, What About "Deadly Plutonium"? Position statement adopted March 1993, WWW.hps.org.
3. M. Walchuk, Human Encounters with Plutonium, Editorial, Health Physics Society's Newsletter, April 1999.
4. ICRP 48, The Metabolism of Plutonium and Related Elements, International Commission on Radiological Protection, ICRP Publication 48, Pergamon Press, 1986.
5. ICRP 26, 30, Recommendations of the International Commission on Radiological Protection and Limits for Intakes of Radionuclides by Workers, ICRP Publications 26, 30, Pergamon Press, 1977 & 79.
6. NCRP 129, Recommended Screening Limits for Contaminated Surface Soil and Review of Factors Relevant to Site-Specific Studies, National Council on Radiation Protection and Measurements, NCRP report No. 129, Issued January 29, 1999, Bethesda, MD, USA.
7. B.W. Church, Dose Assessment Considerations for Remedial Action on Plutonium-Contaminated Soil, Journal of Soil Contamination, Vol.6(3):257-269, 1997.
8. E. Iranzo, and S. Salvador, Potential Radiation Doses to People Living Near Plutonium Contaminated Areas in Palomares, Junta De Energia Nuclear - Spain. Presented at the 29th Annual Meeting of the Health Physics Society, New Orleans, June 3-8, 1984.
9. E. Iranzo, S. Salvador and C.E. Iranzo, Air Concentrations of Pu-239 & Pu-240 and Potential Radiation Doses to Persons Living Near Pu-Contaminated Areas in Palomares, Spain, Health Physics Vol. 52(4), pp. 453-461,1987.
10. E. Iranzo, A. Espinosa and J. Martinez, Resuspension in the Palomares Area of Spain: A Summary of Experimental Studies, J. of Aerosol Sci., Vol. 25(5), pp. 833-841, 1994.
11. E. Iranzo, A. Espinosa and C.E. Iranzo, Dose Estimation by Bioassay for Population Involved in an Accident with Plutonium Release, Second Conference on Radiation Protection and Dosimetry, Orlando, Florida, Nov., 1988.
12. E. Iranzo, A. Espinosa and C.E. Iranzo, Evaluation of Remedial Actions Taken in Agricultural Area Contaminated by Transuranics, IV International Symposium of Radioecology , The Impact of Nuclear Origin Accidents on Environment, Cadarache, France, March, 1988.
13. US DOE/NV-399, A Cost/Risk/Benefit Analysis of Alternative Cleanup Requirements for Plutonium-Contaminated Soils On and Near the Nevada Test Site, May 1995.
14. USDOE/EA-1136, Environmental Assessment for Double Tracks Test Site ,1996.
15. M. Sanchez, A Criteria for the Cleanup of Double Tracks. Personal Communication, USDOE, Nevada Operations Office August, 1996.
16. C. Yu, Et al. A Manual for Implementing Residual Radioactive Material Guidelines Using RESRAD, Version 5.0. ANL/EAD/LD-2 Argonne National Laboratory 1993.
17. USDOE, A Methodology for Calculating Guideline Concentrations for Safety Shot Sites, U.S. Department of Energy, Nevada Operations Office, USDOE/NV--481, June, 1997.
18. Community Advisory Board, Meeting Summary of the Community Advisory Board (CAB) for Nevada Test Site Programmes, Oct. 1, 1997.
19. J.L. Symonds, A History of British Atomic Tests in Australia. Department of Resources and Energy, Australian Government Publishing Service 1985.
20. Technical Assessment Group, (TAG) Rehabilitation of Former Nuclear Test Sites in Australia, Department of Primary Industries and Energy, Australian Government Publishing Service 1990.
21. Department of Primary Industries and Energy, The Maralinga Rehabilitation Project. DPIE Corporate Public Relations, Issue No 3, Commonwealth of Australia 1996.

22. IAEA, International Atomic Energy Agency, Basic Safety Standards, Vienna 1996.
23. G.A. Williams, Editor, Inhalation Hazard Assessment at Maralinga and Emu, Australian Radiation Laboratory, ARL/TR087, May 1990.
24. P.N. Johnston, K.H. Lokan and G.A. Williams; Inhalation Doses For Aboriginal People Reoccupying Former Nuclear Weapons Testing Ranges in South Australia, Health Physics Vol. 63 (6), pp. 631-640, Dec. 1992.
25. J.H. Shinn, Status of Maralinga Resuspension Studies, Letter Report, July 10, 1997.
26. S.M. Haywood and J.G. Smith, Assessment of Potential Doses at the Maralinga and Emu Test Sites; Health Physics Vol. 63 (6), pp. 624-629.
27. K. Palmer and M. Brady; The Diet and Lifestyle of Aborigines in the Maralinga Region, a report prepared for the Maralinga Tjarutua as part of the Maralinga rehabilitation studies, 1988.
28. J.L. Bowen, Ambient Airborne Particulate Matter and Meteorological Data, Clean Slate 1 Post-Corrective Action, Tonopah Test Range, Desert Research Institute Document Number 6358-683-8560.4D1, August, 1998.
29. J.H. Shinn, S.E. Patton and F.J. Gouveia, Seasonal Variations and Frequency Distributions of Plutonium Resuspension at Tonopah, Nevada, UCRL-JC-126286, Lawrence Livermore National Laboratory, April 1997.
30. J.H. Shinn, Range of Resuspension Factors from Contaminated Sites around the World, Personal Communication, April 1999.
31. J.H. Shinn, Complementary Pu Resuspension Study at Palomares, Spain (In Draft), Livermore National Laboratory; February 1999
32. U.S. Dept. of Energy, Double Tracks Test Site Closure Report, Environmental Restoration Division, USDOE/NV-11718-113, June 1997.
33. J.L. Bowen, Air Quality and Meteorological Measurements after Excavation at Double Tracks, TTR, Desert Research Institute Document Number 6357-683-7567.1D1, June, 1997.
34. P.N. Johnston, G.A. Williams, P.A. Burns and M.B. Cooper, Plutonium Resuspension and Airborne Dust Loadings in the Desert Environment of Maralinga, South Australia, J. Environmental Radioactivity 20 (1993) pp. 117-131.
35. J.H. Shinn & F.J. Gouveia, The "Footprint" Area Influencing a High Volume Air Sampler, Lawrence Livermore National Laboratory report No. UCRL-ID-112181, October 1992.
36. S.R. Riedhauser, Verification of KIWI Data from TTR Sites, Bechtel Nevada Interoffice Memorandum, March 8, 1999.

# DATABASE MANAGEMENT OF RADIOECOLOGICAL DATA FOR DOSE ASSESSMENT

I.V. KHROMUSHIN, D.K. DAUKEEV
*National Nuclear Centre of the Republic of Kazakhstan,*
*Almaty, Kazakhstan*

C.F.V. MASON, B. SCHEBER
*Nuclear Materials Technology Division,*
*Los Alamos National Laboratory,*
*Los Alamos, NM, 87545 USA*

There are several sources of man-made radioactivity in Kazakhstan, including uranium ore mining and processing as well as the consequences of nuclear bomb explosions. The issue of radioactive contamination of the Semipalatinsk Nuclear Test Site (SNTS) has been discussed. Our task now is to evaluate the health risk to people from these kinds of contamination.

This problem is very complex and its solution depends on many different factors, such as the radioactive sources, the environment, and the radionuclide behaviour in the environment, etc. There is great amount of world-wide experience in dealing with different aspects of this problem. Consequently, there is a desire to involve the international scientific community in making decisions regarding the radioecological problems of Kazakhstan.

The best way to ensure international collaboration is through direct contact between scientists at different seminars and conferences such as this one. But it is impossible to do this often and attract the world-wide scientific community for exchange of expertise. For this reason, we suggested an idea of virtual collaboration on the Internet between scientists for making decisions regarding the radioecological problems of Kazakhstan. This idea was supported by the United Nations Educational, Scientific, and Cultural Organisation (UNESCO) and the Los Alamos National Laboratory. As a result of these joint efforts, we managed to set up a virtual laboratory (VL) on radioecological research. This laboratory is open for participation by any interested partners. So that as many partners as possible are involved, a laboratory home page with related information and an interactive registration form was prepared.

The core of the VL is a radioecological database that includes data on radiation contamination of SNTS by Cs and Sr, as well as information on the properties of soil,



*S.S. Hecker et al. (eds.), Nuclear Physical Methods in Radioecological Investigations of Nuclear Test Sites*, 221-222.
© 2000 *Kluwer Academic Publishers. Printed in the Netherlands.*

rock, water, etc. Microsoft Access and Dbase were used to create this database. A specific feature of the database is its availability by means of interactive forms for remote partners through the Internet (Fig. 1). To make this database available for remote partners, we are working on writing common gateway interface (CGI) scripts. As a tool for creating a CGI, we have used the Perl programming language because it is rather flexible and does allow for the manipulation of data very effectively. The advantage of this approach is the possibility for partners to query, add, and edit data by using a simple Internet browser.

Java scripts are used as an interface. A service such as Telnet will be used for remote data processing. Of course it is also necessary to discuss such services as ftp and electronic mail that are used for data exchange and communication.

We believe that virtual collaboration makes it possible to solve the radioecological problems of Kazakhstan more effectively by attracting scientists from around the world to offer their solutions to radioecological problems of Kazakhstan.

If individual scientists want to join the VL or to make a contribution to the development of any part of the project, they are welcome to do so.

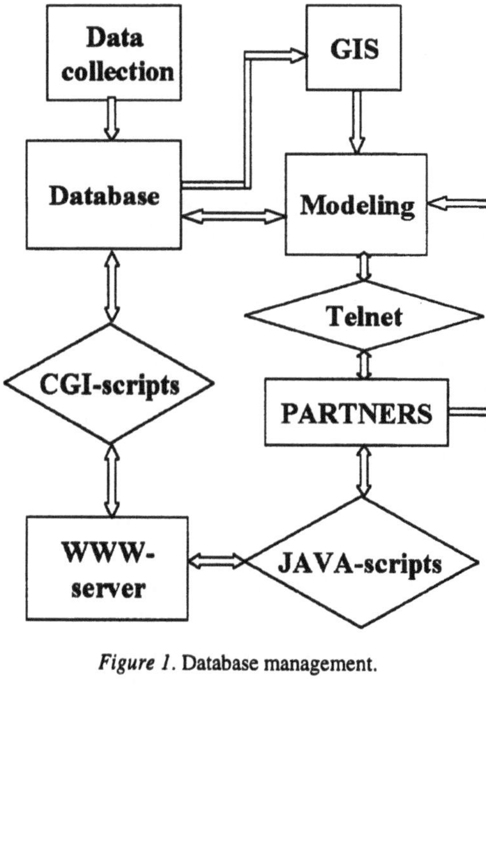


*Figure 1.* Database management.

# DATABASE OF THE RADIOECOLOGICAL SITUATION AT THE SEMIPALATINSK NUCLEAR TEST SITE

K.K. KADYRZHANOV, S.B. KISLITSIN, K.M. KIKKARIN,
A.T. KUKETAEV, A.D. LOPUGA, T.E. TURKEBAEV, Kh. SIRAJET
*Institute of Nuclear Physics,*
*National Nuclear Centre of the Republic of Kazakhstan,*
*Almaty, Kazakhstan*

## Abstract

A database allowing automatic access to available and arriving data on the radioecological situation in the territory of the former Semipalatinsk Nuclear Test Site (SNTS) has been designed at the Institute of Nuclear Physics of the National Nuclear Centre of the Republic of Kazakhstan. This database will make it possible to accumulate, systematise, and computerise data on the radioactive contamination of this territory. This paper presents the current condition of the database. This database should not be considered as a final system since new data and capabilities for data analysis will be modified and optimised.

## 1. Background

One of most significant problems of the National Nuclear Centre of the Republic of Kazakhstan founded in 1994 is the detailed clearing up of the radioecological situation as well as the constant monitoring and elimination of consequences of nuclear-weapon tests at the Semipalatinsk Nuclear Test Site (SNTS). Carrying out the research on the SNTS territory contaminated by radioactive materials and fulfilling rehabilitation measures have lead to the creation of a computer database (DB) on the radioecological situation at the test site territory. This database will make it possible to treat and systematise the available data on the ecological conditions in this territory and to add new data on the radionuclide contamination there. The DB will also allow researchers to estimate the impact of various nuclear test characteristics on the environment and on the health of the population of the SNTS region.

To accomplish these analyses, the Institute of Nuclear Physics of the National Nuclear Centre organised a local computer network as described below.



*S.S. Hecker et al. (eds.), Nuclear Physical Methods in Radioecological Investigations of Nuclear Test Sites*, 223-230.
© 2000 *Kluwer Academic Publishers. Printed in the Netherlands.*

## 2. Database: Characteristics and Types

### 2.1. HARDWARE AND SOFTWARE

The hardware and the software of this network make possible the development of databases using client-server technology to provide input on electronic geographical and geological maps and to map radioecological data on these electronic maps.

The local area network sets and is configured as the domain of the central server. The network server operates under Windows NT Server 4.0 (TCP/IP and NetBEUI protocols); client workstations are under operating systems Windows 95 and Windows NT Workstation 4.0 (TCP/IP and NetBEUI protocols). In total, within the network there are seven workstations, equipment for operating with the Geological Information System (GIS) - digitizer and scanner, and a plotter for printing geographical maps (format A1).

For DB development, GIS and additional handlers of the data from the DB, the software Oracle 7 Workgroup Server (release 7.3 for Windows NT, version 7.3.3.0.0), Microsoft Access, Delphi 3.0, C ++, PC ArcInfo 3.51, ArcView 3.0a will be used.

### 2.2. STRUCTURE, CONTENT, AND CAPABILITIES OF THE DATABASE

The structurally functional scheme of the DB is illustrated in Figure 1. As this figure shows, separate storage of the cartographic information in the cartographic database (CDB) and data on the $\alpha$, $\beta$, $\gamma$ background, the contents of radionuclides in sampling points, and the nuclear explosion data in the radiological database (RDB) are all stipulated. Such a system arrangement allows independent RDB and CDB operations and makes it possible to use other types of cartographic software and, if necessary, to operate without cartographic visualisation.

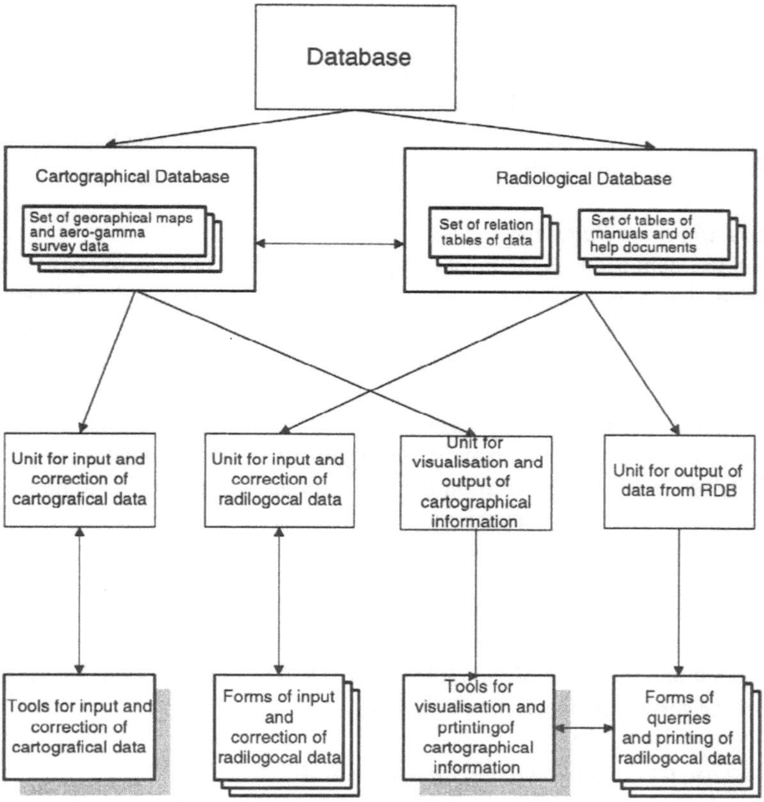


*Figure 1*. The structurally functional scheme of database on radioecological situation on SNTC

The RDB consists of a set of relational data tables and manuals and contains data accessible for measuring α, β, γ background as well as the content of radioisotopes in soil, water, etc., distributed under different tables. Each record contains the date of sampling, the coordinates, and the description of the sampling point (lake, river, tunnel, etc.). In addition, as seen in Figure 1, the tools for information input and editing (i.e., input forms in the data tables and manuals), tools for creating reports and export of data in CDB are also designed.

This CDB stores the cartographic information about geographical objects and sampling points. If necessary, part of the ecological data can also be duplicated in the RDB. The CDB also includes the geographical maps of a scale of 1:100000 and 1:200000, as well as aero-gamma survey data having topographical control and attributive information on geographical objects. The measurement data of radionuclide concentrations in the sampling points and the α, β, γ background, etc. are exported from the RDB into the GIS and then drawn on a geographical map by particular color of points or isolines relevant to a particular level of concentration of a radionuclide or α, β, γ background.

## 2.3. RADIOECOLOGICAL DATABASE: CHARACTERISTICS

The RDB consists of two parts that are practically independent from each other: the nuclear explosions database and the database on the radionuclide contamination of the test site territory. There is also a set of normative and manual documentation in the PDF format (for reading with Acrobat Reader).

### *2.3.1. Test Explosions Database*

The DB on test explosions is the most complete because no nuclear weapons tests have been carried out in this territory since 1992. This DB contains the following data distributed under several tables: date of explosion; time of explosion; explosion power; explosion site; coordinates of the point of explosion; the type of explosion, etc.

### *2.3.2.Radinuclide Contamination Database.*

The DB on radionuclide contamination consists of two parts –a database on a surface radioactivity of Cs-137, obtained from the results of an aero-gamma automobile survey and one from radionuclide concentration in soil data received as a result of a probe drawing in sampling points and of subsequent laboratory measuring.

### *2.3.3.Database on the Cs-137 Surface Radioactivity*

This DB contains the table presenting the information about the coordinates of the sampling point and the value of the radioactivity ($Ki/km^2$) in this point. Figure 2 shows the map of the surface radioactivity of Cs-137.

### *2.3.4. Database on the Radionuclide Content in Soil*

This DB includes a set of tables, input forms, tools for exporting data from the RDB into the CDB, and tools for obtaining the reports. This part of the database contains the following information: the code of the sampling point; the date of the probe drawing; the coordinates of the sampling point; the method for determining radionuclide content.

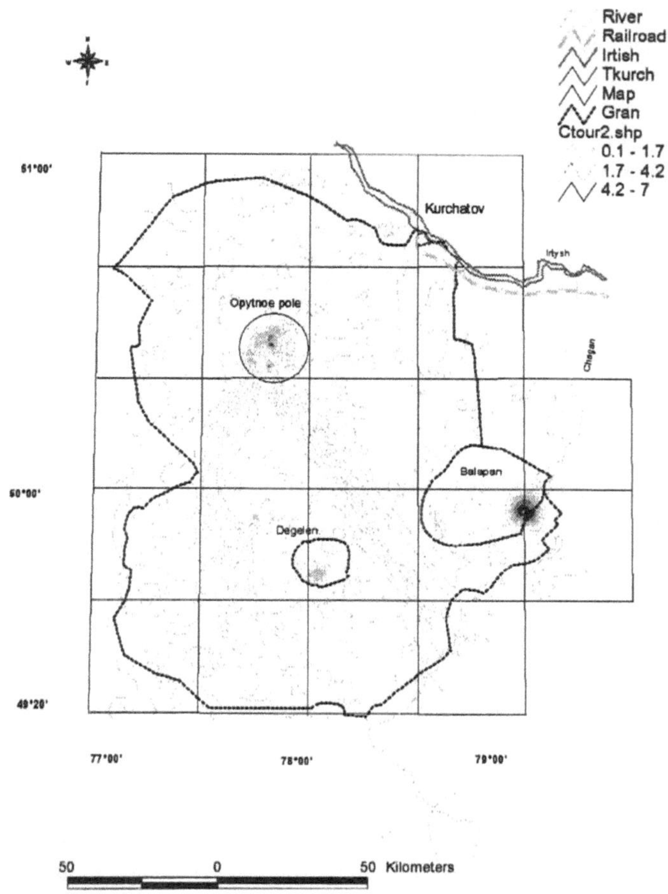


*Figure 2.* Isolines of surface activity of Cs-137 mapped from data of the aero-gamma survey of 1994, and the automobile and foot gamma survey of 1997

## 2.4. THE CARTOGRAPHICAL DATABASE

The cartological database (CDB) is intended for storage of the electronic geographical information together with the attributes of the objects and the visualisation of the data from the RDB by the plotting of the contaminated sites on the electronic geographical map. The development of the CDB was carried out by means of GIS: PC ArcInfo (version 3.51) and ArcView. With the help of the software packages ArcInfo, input of the geographical information and attributes of geographical objects was conducted: This information includes the contour of the terrain, the lakes, river, roads, and occupied points. A map of the Cs-137 surface contamination of the SNTS territory with data from the aero-gamma survey (obtained in 1994, and improved by the data of the automobile and foot gamma survey in 1997 at separate territories of test sites ("Opytnoe Pole," "Balapan," and "Murzhik")) was entered separately. Various procedures of information input were applied with the help of a digitiser (grains from the test site, basins, river, road) and also by tracing the images of topographical maps obtained by scanning and using other data input tools of ArcInfo - editor Data Conversion from text files (surface contamination by caesium).

On the items listed above, the layers (contour of the test site, roads, lakes, etc.) that are activated are those that can be mapped all together in one document or those that are indispensable at the present moment. Figure 2 presents a map of the Cs-137 surface contamination of the SNTS territory. This map shows that the Cs-137 surface activity is distributed nonuniformly throughout the test site territory. It is possible to mark the separate contaminated sites relevant to tracks of radionuclide contamination from the most powerful atmospheric and underground nuclear weapon tests. So tracks of radioactive fallout at the southeast and southwest sites, i.e., the Opytnoe Pole, and at others where the surface Cs-137 activity is significant, are clearly visible. At the same time, there are sites in the SNTS territory where the surface activity of caesium does not exceed a level of Cs-137 global fallout for this terrain.

To visualise the radionuclide contamination information from the RDB on a geographical map, data export from the KDB is applied. The data record for the visualisation export is contained in separate files in a special folder "ArcView Export," which contains information for the KDB. With the help of ArcView, these data can be viewed as new layers on geographical maps.

An example of the visualisation of the plutonium concentration measured in the SNTS territory is shown in Figure 3. The open circles indicate that the values of the Pu-239 and Pu-240 concentrations in that sampling point are less than 400 Bq/kg; the dark circles mean that the Pu-239 and Pu-240 concentrations are great than 400 Bq/kg. The simultaneous drawing of different layers makes it possible to analyse the different nuclide allocations on a test site territory and to make inferences about a correlation in the content of different radionuclides at one sampling point.

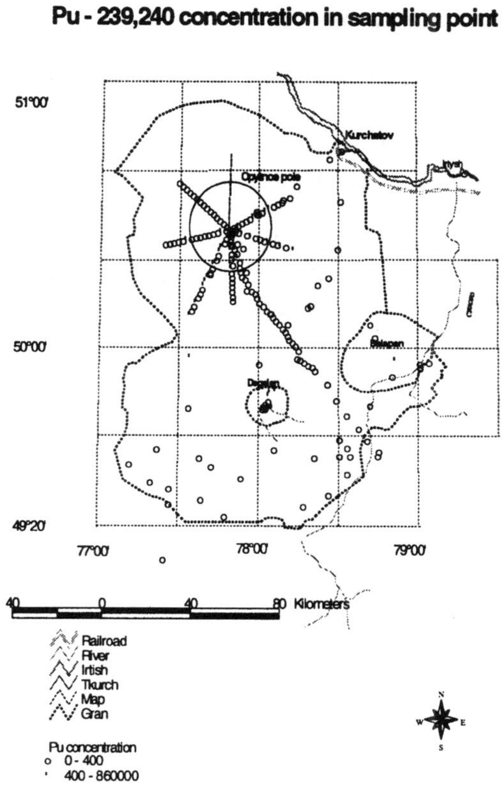


*Figure 3.* Concentration of Pu-239, 240 in a sampling point in the SNTS territory

## 3. The Reports

One part of the DB makes it possible to access the same reports, having picked the necessary parameters. At the present moment, the following reports are accessible:

***All nuclear explosions***—scanning of the list of explosions. This report contains all accessible data on nuclear explosions.

***The radionuclide content in a given square***—The user sets geographical coordinates of the initial and last points of the square and type of radionuclide (along longitude and latitude) and gets all data about the soil concentration of a certain radionuclide in a given square.

***All sampling points in the band***—The user needs to set the initial and last points of the band and width of the band in geographical degrees of the Earth's surface) In this report all information about the sampling point in a band of a given width is available.

***Nuclear explosions at a given site***—The user picks the site of interest from the list and receives the list of nuclear explosions that were conducted on that site.

***The list of sampling points where the content of the nuclide is higher than a given level*** - The user picks the radionuclide of interest from the list, sets a level of radionuclide concentration, and receives the report containing the list of sampling points where the concentration of certain radionuclides exceeds a given level.

***Allocation of nuclear explosions in specific years***—The user sets the initial year N and an interval M in years and receives the report about allocations of the number of test explosions yielded in each given M year interval, since N that year. The example of the report, ***The Radionuclide Content in a Given Square,*** is listed in Table 1.

TABLE 1. The radionuclide content in a given square [Interval of the attitudes: 50-50.5 Interval of the longitudes: 77-77.5]

| **Nuclide** | **CS-137** | | | | | |
|---|---|---|---|---|---|---|
| **ID** | **Value** | **Unit.** | **Error %** | **Altitude** | **Longitude** | **Remark** |
| W-12 | 34 | Bq/kg | 25.00 | 50.41 | 77.48 | |
| W-13 | 21 | Bq/kg | 9.00 | 50.41 | 77.45 | |
| W-14 | 40 | Bq/kg | 6.00 | 50.40 | 77.42 | |
| W-15 | 9 | Bq/kg | 10.00 | 50.40 | 77.40 | |
| **Nuclide** | **K-40** | | | | | |
| **ID** | **Value** | **Unit.** | **Error %** | **Altitude** | **Longitude** | **Remark** |
| •-12 | 768 | Bq/kg | 10.00 | 50.41 | 77.48 | |
| •-13 | 556 | Bq/kg | 8.00 | 50.41 | 77.45 | |
| •-14 | 434 | Bq/kg | 10.00 | 50.40 | 77.42 | |
| •-15 | 529 | Bq/kg | 15.00 | 50.40 | 77.40 | |
| **Nuclide** | **Th-232** | | | | | |
| **ID** | **Value** | **Unit.** | **Error %** | **Altitude** | **Longitude** | **Remark** |
| •-12 | 22 | Bq/kg | 25.00 | 50.41 | 77.48 | |
| •-13 | 15 | Bq/kg | 9.00 | 50.41 | 77.45 | |
| •-14 | 14 | Bq/kg | 12.00 | 50.40 | 77.42 | |
| •-15 | 26 | Bq/kg | 30.00 | 50.40 | 77.40 | |
| **Nuclide** | **U-238** | | | | | |
| **ID** | **Value** | **Unit.** | **Error %** | **Altitude** | **Longitude** | **Remark** |
| •-12 | 34 | Bq/kg | 20.00 | 50.41 | 77.48 | |

## 4. Conclusion

In summary, it is necessary to note that the DB on ecological conditions in the SNTS territory as presented in this paper can not be considered as the final system once the new data structure and the content of the database are finished and optimised. The new DB will make it possible to examine new aspects of the contamination of the SNTS territory (the examination of air basin, migration of radionuclides). The DB will also permit new opportunities and building blocks.

# MODERN METHODOLOGY AND EQUIPMENT FOR FIELD RADIOMETRY AND SPECTROMETRY

# METHODS AND EQUIPMENT FOR FIELD MONITORING OF DIFFERENT TYPES OF $^{137}$CS CONTAMINATION, INCLUDING SMALL, ISOLATED SPOTS

O. SLÁVIK
*VU'JE Research Institute,*
*Trnava, Slovakia*

**Abstract**

The 18-km-long banks of the Manivier Canal and the Dudvah River, the recipients of the Bohunice Nuclear Power Plant (NPP) waste water, are contaminated by $^{137}$Cs as a result of two plant accidents in 1976 and 1977. In total, over 76 000 m$^2$ of these banks (the lower and central parts) have been found to be contaminated to levels exceeding 1 Bq$^{137}$Cs/g of soil. This paper describes parts of the planning for restoring these banks and emphasises the field monitoring techniques used for site characterisation.

## 1. Introduction

The significance of the contamination of the river and canal banks was discovered in 1991 in conjunction with implementation of a flood-control project on the small Dudvah River. The 18-km-long banks of the Bohunice Nuclear Power Plant (NPP) waste water recipients, the paved Manivier Canal and the Dudvah River, are contaminated by $^{137}$Cs as a result of two NPP accidents in 1976 and 1977. In total, over 76 000 m$^2$ of these banks (the lower and central parts) have been found to be contaminated to levels exceeding 1 Bq $^{137}$Cs/kg of soil.

Soon after the discovery of the $^{137}$Cs contamination problem, proper post-accident restoration actions were requested. The cleanup level 1 Bq/g of $^{137}$Cs was authorised for this purpose. So in 1992, as part of the bank restoration project, field monitoring of the scale and boundaries of the contamination was initiated by the NPP as the first step in carrying out proper remedial measures. It was the moving of large amount of the contaminated soil from the river banks, and its release into the environment during the flood control project implementation that posed the most serious potential threat for the nearby population.

Since 1993, the VU'JE Research Institute has been involved addressing, comprehensively, the contamination problems through a radiological survey and environmental-restoration planning. This paper describes parts of the planning for the



*S.S. Hecker et al. (eds.), Nuclear Physical Methods in Radioecological Investigations of Nuclear Test Sites*, 233-237.
© 2000 *Kluwer Academic Publishers. Printed in the Netherlands.*

bank restoration and emphasises the field-monitoring techniques used for the site-characterisation purposes.

## 2. Contamination at the Site

Between 1976 and 1978, when two accidents on the $CO_2$-cooled and heavy-water moderated NPP-A1 unit occurred, construction of a flood-control project on the Dudváh River was just being implemented in an 8-km-long lower section upstream of the river's mouth. In the upper part of the river, narrow stripes of the lower bank surfaces were naturally contaminated during the accidental releases. This contamination occurred because of the water level elevation caused by bypassing the flow 8 km upstream from the mouth during a construction project. At the river's mouth, the lower bank terraces, which were newly constructed under the mentioned flood-control widening project, along about a 3-km-long river section were unexpectedly found to be highly, but unnaturally, contaminated. It is assumed that the highly contaminated bed sediments from the bypassed river bed could be used as a cover for the built terraces.

The geometry of this $^{137}$Cs contamination, in the both first cases, created the following:

1. Narrow strips (slab) in the inner lower bank where the top 15–20-cm-thick soil layer is mainly by 137Cs as a result of about a 20-year downstream dispersion processes.

2. Highly intense small 137Cs contamination spots discovered in a 2-km-long section of the canal banks downstream from the NPP release starting point. This hot spot contamination (on the central part of the bank surfaces) occurred, probably, as a result of the paved canal cleanup after a flood. During this cleanup, highly contaminated mud spread from the contaminated NPP-A1 sewage system, after flooding in 1980, and then was scattered manually from the paved canal bed over the central part of the bank surfaces.

3. A large area of $^{137}$Cs contaminated arable soil that was found in a field alongside the river near an irrigation pump station.

It is obvious that each of these three types of contamination requires a specific measuring and calibration approach for its field monitoring.

## 3. Applied Measurement and Field Monitoring Techniques

Gamma radiation readings and sliced bulk soil samples for laboratory analysis were taken at the surface of the banks inside and outside of the built levees. The laboratory analysis made it possible to determine that $^{137}$Cs is the dominant contaminant in the site. Negligible contents of other radionuclides, such as $^{60}$ Co, $^{134}$Cs, $^{239}$Pu, were also found in the samples. The content of $^{90}$Sr in soil estimated from the values of about 20 analysed samples has been found to be from 50 to 100 times lower than the content of the $^{137}$Cs. Nevertheless, strontium's contribution to effective dose is not insignificant.

For the accessible places in the outer side of levees, continual scanning by a Vehicle Mounted Gamma Survey system (VMGS) was used [1]. The VMGS, which consists of a large shielded scintillation detector (NaI(Tl), 100x100mm), commercially available NIM module electronics, a microcomputer, and custom software, was developed at the VU'JE Research Institute. This technique was used not only for the radiation survey of the 18-km-long affected banks, but, for several hectares of the nearby fields and flood plains, as well.

One contaminated field spread of 2000 $m^2$ along the Dudvah River bank and a limited flood plain area of the Vah River and the former Dudvah River were discovered and evaluated by the VMGS. Consequently, the VMGS was mounted on the hydraulic arm of a tractor and tested also for work on the hardly accessible steep Manivier Canal banks.

Inside the levees, a hand-held, slightly shielded, gamma survey meter with a large plastic scintillator (75x75 mm) was used for discrete grid-point measurements within a 20-m-line grid. A measuring time constant of $\tau$=10s was applied to reach sufficient sensibility of the detector. In critical places with the highest activities, a line grid with a 10-m measuring step was used to achieve about 15% coverage of the surveyed bank surfaces. The shielding of the detector was important for achieving a sufficiently close correlation between the activity concentration and the detector response above the contaminated bank strip of variable widths.

For those canal sections with contaminated spots, continual scanning with $\tau$=1 s to search for the most intense contaminated spots had been carried out in addition to the line grid survey. Some spots were investigated in detail by scanning and soil sampling to determine the shape, size, specific activity, and depth distribution of $^{137}$Cs. There was no close correlation found between the detector response and specific activity of the spots because of the irregular shapes and variable sizes found, which in most cases were smaller than the circle area seen by the detector ($R_{dv}$=70 cm).

Finally, three groups of spots were selected according to the typical average activities determined by laboratory analysis of the samples. Sizes and linear densities of the contaminated spots were roughly estimated on the basis of monitoring device readings. These data were then applied to make conservative assessments of the average contamination of only the lower part affected bank strip.

The detailed and comprehensive survey done between 1991 and 1994 shows that the topsoil contamination on the banks widely varies from background level to 20 Bq/g for the Dudváh River and reaches to 250 Bq/g of $^{137}$Cs for spotty section on the canal banks. The contamination is spread over about a 0.5–3 m wide strip on the lower part of the banks (on average, over 2.2 m) and the average level of $^{137}$Cs in the top 10 cm soil layer reaches 6.3 Bq/g. Thus, the overall contaminated area, with an activity level exceeding 1Bq/g of $^{137}$Cs, is about 67 000 $m^2$ and the volume of soil that had to be removed according to the first cleanup limit exceeded 13 000$m^3$. However, a volume-distribution investigation of the contaminated soil has shown that removing only the mostly contaminated soil would result in a significant improvement in remediation of the contaminated site.

## 4. Monte Carlo Simulations of Plastic Scintillation Detector Response Above Different Circle Slab Sources

A more-accurate evaluation of the Cs 137 contamination discontinuities in the Manivier Canal bank requires the application of more sophisticated measuring and evaluation techniques. It has been shown that one standard measurement reading is insufficient to determine activity concentration in a circular slab source with an unknown radius. This is true because a detector response above a circular soil source generally depends on three independent parameters: the activity concentration, circle radius (or the surface of contamination), and depth distribution. (The last two parameters relate to the total amount of activity in the spot of contamination.) However, field samplings and Cs 137 measurements have shown that typical depth distribution can be well described by an effective homogeneous slab source with 10-cm thickness.

The remaining two parameters have to be assessed according to more than one measurement using e.g. detector responses in two predefined measuring positions. For different shielding of the detector, a ratio of unshielded-to-shielded detector response must be used.

For a simple measurement technique, such as a standard dose-rate meter (i.e., not spectrometric systems), scattered photons play important roles for detecting ionisation gamma-radiation. Therefore, knowledge of complex photon transport through the source and source-to-detector space and detector media, are required to accurately solve the given problem.

In this context, the Monte Carlo simulation of the selected and sufficiently sensitive detector (NB3201 detector with 75x75 mm plastic scintillator) has been carried out by the VU'JE Research Institute and by the Comenius University Nuclear Physics Department in Bratislava. The European Centre for Nuclear Research (CERN) GEANT Monte Carlo code [2] for homogeneous 10-cm-slab circular sources was employed for these purposes.

Typical simulation results have shown that the ratio of two measurements made at 3 cm and 30 cm make it possible to distinguish the diameter of an effective circle source. Though the uncertainty of this type of assessment may be great, it is possible to distinguish the smaller and the larger $^{137}$Cs spots by the described method, at least. This method is the basis for a more satisfactory characterisation of existing spotty contamination on the river and canal banks.

For smaller $^{137}$Cs spots with radii up to 30 to 35 cm, the ratio of two detector response and a detector response in 30 cm measuring height-to-total activity conversion factor (pGy/s per kBq of $^{137}$Cs ) of the total amount of Cs activity in the spot has been found to be the more appropriate measurement and evaluation technique. For larger $^{137}$Cs soil spots, the dose rate-to-activity concentration conversion factor (pGy/s per kBq/kg of $^{137}$Cs) with only slight source radius dependence can be used to estimate the concentration of the contamination spot activity.

## 5. Conclusions

Reevaluation of a $^{137}Cs$-contaminated bank restoration project has been conducted for the Bohunice NPP site with comprehensive and detailed site monitoring techniques. Thanks to the application of a consistent site-characterisation technique and dose-assessment method, unnecessary waste soil disposal will be avoided in the new bank-restoration project. The Monte Carlo simulation was used to propose sophisticated field-monitoring techniques for restoration and postrestoration monitoring on some spotty contaminated area. The method proposed for $^{137}Cs$ field monitoring is based on dose rate measurement technique and on exploring the ratio of two measurement readings for different measuring height above the centre of contaminated spot. The method is currently undergoing final development and verification in real field conditions.

## 6. References

1. SLÁVIK,O., MORÁVEK,J., Identification and Radiological Characterization of Contaminated Sites in the Slovak Republic, Proc. Planning for Environmental Restoration in Central in Eastern Europe" (Budapest, 4 - 8 October 1993), IAEA-TEC-DOC- 865, Vol. 1, 1996
2. GEANT User's guide, CERN Program Library Office, CERN-CN, CH-1211 Geneva23
3. Based on report presented in the Research Agreement No 8731/OR between IAEA and VU'JE, at the IAEA second R C M meeting on Site Characterisation Techniques used in Environmental Restoration, CIEMAT Facility, Madrid 3-7 November 1997

# GEORADIOECOLOGICAL MAPPING PROBLEMS AT NUCLEAR EXPLOSION SITES

D.N. BELYASHOV, YU.A. GRINSHTEIN
*Institute of Geophysical Investigations,*
*National Nuclear Centre of Kazakhstan,*
*Altmay, Kazakhstan*

I.M. KHAIKOVICH
*Research Institute of Prospecting Geophysics (RIPG),*
*St. Petersburg, Russia*

## 1. Introduction

Since 1946, large-scale, weapons of mass destruction tests have taken place in Kazakhstan. The Semipalatinsk Nuclear Test Site (SNTS) is the area where the highest number of nuclear weapon tests took place; about 470 nuclear explosions were conducted there over a period of 40 years.

Recently, the following aerial gamma-spectrometric surveys of different scales from 1:300 000 to 1:25 000 have been conducted at SNTS: the automated gamma-spectrometric survey on a scale of 1:50 000, the lithogeochemical survey on a scale of 1:50 000, the pedestrian spectrometric survey with assay sampling, and other less significant investigations.

Figures 1 and 2 show the maps of $^{137}$Cs distributions plotted from the aerial geophysical survey of the Balapan site on scales of 1:300 000 and 1:25 000 correspondingly. While comparing the data in these figures, one can see that certain local radiative fields of high intensity coinciding with combat borehole mouths have been missed by the small-scale survey.

To correctly substantiate the sampling network parameters of these areas, a range of lithochemical profiles, including the profile of a 10-km-long crossing of the combat boreholes (1061, 1315, 1304, and Glubokaya) was covered in a local plot of the Balapan site. Traverse profiles that were 1-km-long were covered through profiles' crossing the mentioned boreholes. Assays were sampled at an interval of 50 meters. Sampling was conducted at depths of 5 cm and 20 cm. There was a 2-kg-mass probe, with 560 probes being sampled. The sampling scheme is presented in Figure 3.



*S.S. Hecker et al. (eds.), Nuclear Physical Methods in Radioecological Investigations of Nuclear Test Sites,* 239-244.
© 2000 *Kluwer Academic Publishers. Printed in the Netherlands.*

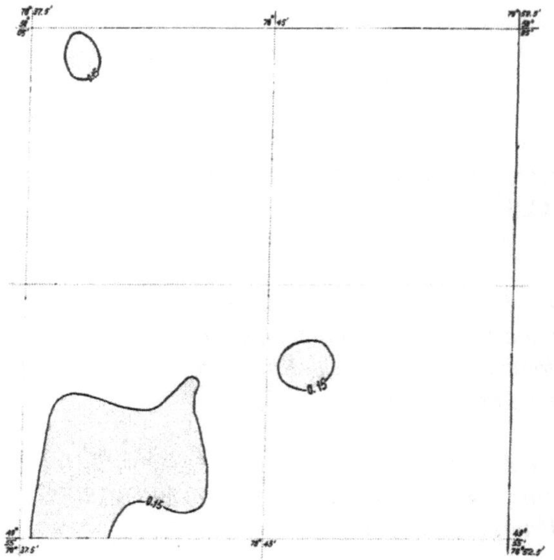

*Figure 1.* A map fragment of the density of SNTS's $^{137}Cs$ ground contamination isolines (according of the data of an aerial gamma-spectrometric survey on a scale of 1:300 000).

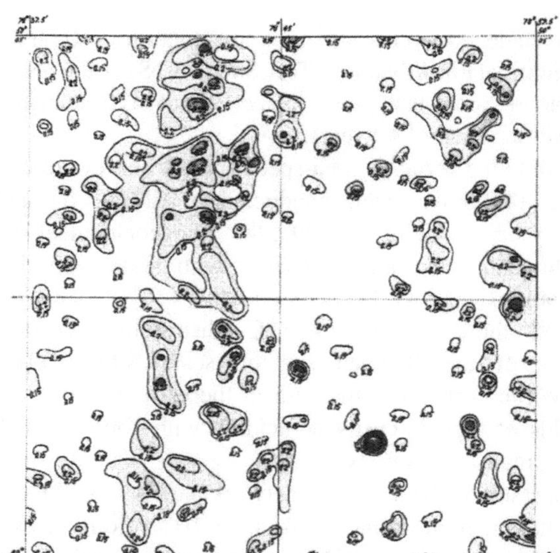

*Figure 2.* The map of the density of Balapan's $^{137}Cs$ ground contamination isolines (according to the data of an aerial gamma-spectrometric survey on a scale of 1:25 000).

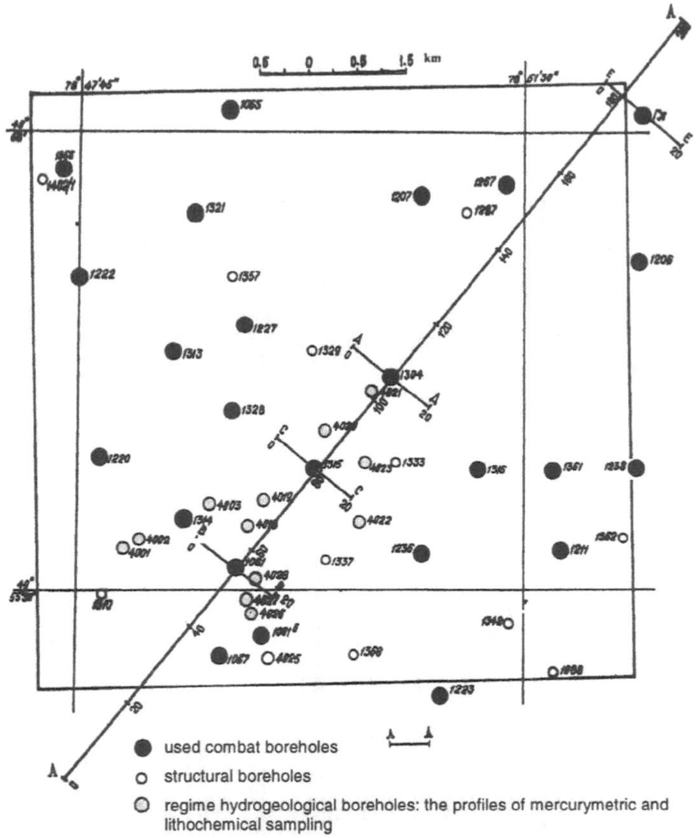


*Figure 3*. Scheme of object allocation within the local plot.

## 2. Assay Analysis

Assay analysis for radionuclides was carried out with the Institute of Nuclear Physics at the National Nuclear Centre of the Republic of Kazakhstan. Figures 4 and 5 show the results of sampled probe analyses for $^{137}Cs$ and $^{241}Am$, conducted at a depth of 20 cm. These figures show the radionuclide distribution along the profile to be extremely heterogeneous, while the anomaly size does not exceed 100-150 meters.

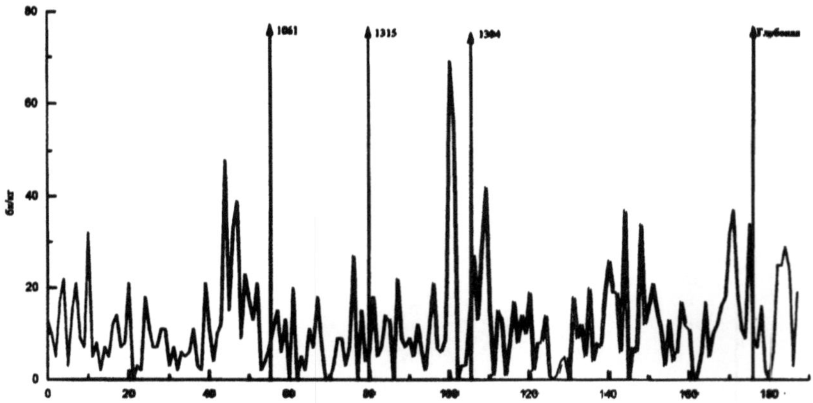

*Figure 4*. Results of analyses for $^{137}$Cs of the Balapan site, the profile 12 at a depth of 20 cm.

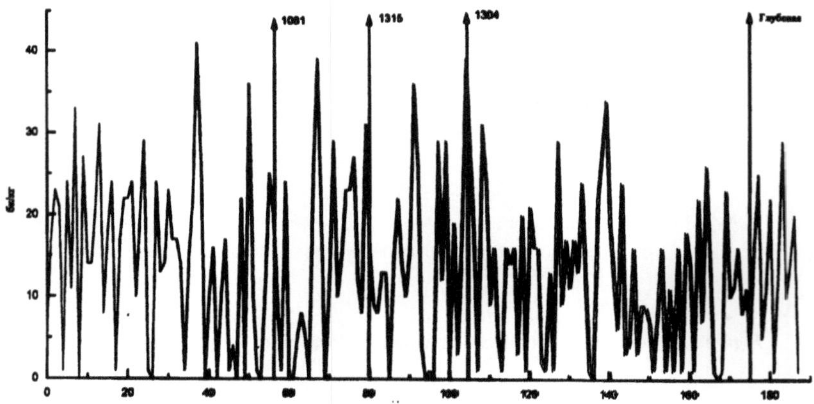

*Figure 5*. Results of analyses for $^{241}$Am of the Balapan site, the profile 12 at a depth of 20 cm.

## 2.1. RESULTS

Analysis of the sampling work makes it possible to allow the following inferences to be made:

1. The main specific feature of radiation fields in the SNTS territory is their extreme heterogeneity, which gives them a mosaic or dappled structure. Consequently, it is impossible to predict the character of radionuclide distribution in the locations where small-scale aerial gamma-spectrometric surveys have not revealed fields of anomalous contamination;

2. Correct radiation and ecological zoning of the SNTS territory for selecting safe areas for economic activity can be accomplished with the help of geophysical survey on a scale not less then 1:25 000. This zoning will provide an estimate of the distribution regularities of the exposure rate of external gamma radiation for both technogenic ($^{60}$Co, $^{90}$Sr, $^{137}$Cs, $^{(238,\ 239+240)}$Pu) and natural (U, Th, K) radionuclides;
3. If traditional methods of gamma spectrometry are used, the problem of estimating territory contamination with such extreme dangerous technogenic radionuclides as $^{90}$Sr and $^{239}$Pu is still not resolved. It needs to be solved with the help of a ground lithochemical survey on the scale of 1:25 000 to 1:50 000 only. This type of measurement corresponds to from 160 to 40 assay samplings from one km$^2$. It is necessary to sample and analyse about 700 000 assays, given that the test site area is about 18 000 km$^2$.

The capacity of Kazakhstan's laboratories certified to make analyses for such radionuclides is not more then 3000 assays annually. Hence, several centuries of work and tremendous investments would be required to solve the task by lithochemical survey based on the regular network. However, the current cost of probe analysis for Pu content ranges from U.S. $500 to U.S. $1000. That more reasonable cost makes such area sampling of the territory now possible. Determining the contamination level and structure of the studied territory can now be accomplished with a more cost-effective use of time and resources.

## 2.2. NEW METHODS

Complex investigations should be carried out to solve the task of radiocontamination assessment in the entire SNTS territory by using aerial and ground gamma spectrometry and geochemical sampling. The theories and methods of studying radioactive contamination distribution through aerial and ground gamma-spectrometric techniques are explicitly described in the literature. However, those areas contaminated with alpha- and beta-radioactive nuclei ($^{90}$Sr, $^{239\text{-}240}$Pu) can not be investigated by these particular methods.

The searching and prospecting carried out at uranium and other ore deposits show that authentic information on radionuclide distribution can be obtained with the help of geochemical survey based not only on the regular network but also on the dispersion trains as well, through assays being sampled from the bed deposits.

However, the lack of experience in carrying out this type of lithochemical survey of dispersion patterns under the SNTS conditions (dry climate, lack of constant water supply, appearance of occasional water from melting snow and rare rains, fine-hilly and even plane relief) requires adapting the sampling technique to meet the conditions of the region under investigation.

In addition to the sampling technique, a field procedure of proximate analysis for $^{90}$Sr and $^{239\text{-}240}$Pu must also be designed and the equipment and operating methods developed. This procedure is certain to provide simultaneous analysis for the gamma-radiative nuclides $^{137}$Cs, $^{60}$Co, as well as for U(Ra), Th, and K.

In this connection, it seems reasonable to design the analytical procedure of the lithochemical sampling on the basis of the temporary water supply and to create the field instrumental and methodical system for prompt assay analysis of the most toxic alpha-, beta- and gamma-radiative nuclides $^{137}Cs$, $^{60}Co$, $^{90}Sr$, $^{239\text{-}241}Pu$.

Determining Pu content will be carried out with the help of the alpha-spectrometric method after a brief radiochemical pretreatment (radionuclide concentration without separation). An ionisation chamber 0.25 $m^2$ of active area will be used as the detector. Compared with the cost of traditional radiochemical methods, this approach significantly simplifies and reduces the price of sample chemical pretreatment. In addition, the productivity of an equipment set will come to from 5 to 20 assays per working shift (depending on the contamination level).

Using the method of beta-radiation spectrometry, the same field laboratory will determine Sr content. The method of beta-spectra decomposition into the system of orthogonal images will be used and will provide reliable determination of $^{90}Sr$ content in the assays with the specific activity beginning from 30 Bq per kg measured against the background of $^{137}Cs$ and natural radionuclides.

These analytical procedures of spectrometric determination of Cs and other gamma-radioactive nuclei are well known and provide high productivity at low cost. The expected productivity of this laboratory design ranges from 2000 to 4000 analyses per year.

The analytical procedure of lithochemical search based on dispersion patterns will consist of determining the sampling fractions, the methods of sample enrichment, and the density of sampling under various field conditions.

Thus, because the analytical procedure of lithochemical sampling based on dispersion patterns can correctly map radiocontamination fields with the minimum of assays sampled, this method is being considered for development and trial under the SNTS conditions. This analytical tool makes it possible to select within the shortest time the areas that can be used economic activity.

In addition, the question of the fitness of the technique, the field instruments, and methods to map radionuclide contamination in the SNTS territory and within the numerous test sites in the Republic of Kazakhstan and abroad is also to be considered.

# ACTIVATION ANALYSIS OF LARGE VOLUME SOIL EXTRACT SAMPLES

R. GWOZDZ
*Tracechem, Markmandsgade 2, 3 th, 2300 Copenhagen S, Denmark*

F. GRASS
*Atominstitut, Schuettelstr. 115, 1020 Vienna, Austria*

G. POLIKARPOV
*Faculty of Comparative Radioecology*
*Institute of Biology of the Southern Seas,*
*Prosp. Nachimova, 2, 335011 Sevastopol, Ukraine*

YU. KUTLAKHMEDOV
*Biological Faculty, Kiev University, Vladimirskaya Str. 64,*
*252601 Kiev, Ukraine*

## Abstract

In the process of transferring stable and radioactive isotopes from soil to plants, a soil solution is the intermediate phase. A transfer coefficient from soil to plants can be divided into two separate distribution coefficients—one between a soil and a soil solution and the other between a soil solution and a plant. A new method, the so-called large-sample activation analysis, has been applied to determine stable species in plants and extractant solutions, simulating soil solutions.

## 1. Introduction

Agricultural areas that have been contaminated by long-lived activity, and especially by caesium-137, have to be carefully monitored for many years. The strong affinity of the caesium ion to clay minerals is one of the causes of the very strong time-dependent binding of caesium in soils. This binding reduces the ability of radioactive caesium to be transferred to plants and, in more general terms, reduces the caesium release mobility into the biosphere. Only the part of caesium present in the soil—in mobile form—can be transferred to plants. This mobile part is dependent on many parameters characterising the soil, as well as on the soil solution system, and on the method by which radioactive caesium was introduced to the soil. If caesium is introduced to soils



*S.S. Hecker et al. (eds.), Nuclear Physical Methods in Radioecological Investigations of Nuclear Test Sites*, 245-252.
© 2000 *Kluwer Academic Publishers. Printed in the Netherlands.*

as a result of radiation accidents, field experiments, lysimeter or pot experiments, its binding into a nonexchangeable form grows with time. For instance to compare the data originating from field experiments to those from artificial contamination, a correction coefficient for the transfer factor is proposed [1]. Between 1 and 2 years after the experiment, no correction is necessary; under 1 year this factor should be 2.5, and after from 2 to 5 years, it would fall to 0.5. For a time interval of 5 to 7 years after the experiment, the correction factor would be further reduced to 0.1, and after 7 to 12 years to 0.05.

Many other parameters strongly influence the transfer from soil to plants: soil type, clay minerals in soil, organic contents in soil, mycorrization of the root system, genotype, and the elemental composition of a soil. A strong interrelation exists between the stable elements in the soil and the radioactive isotopes of the same elements introduced artificially into the soil [2–5]. There is also an interrelation between chemically close elements, such as barium and radium, or rare earth elements and americium.

With the use of a weak extractant, for instance 1M ammonium acetate, the mobile fraction of radioactive caesium, which could undergo a transfer to plants, is presumed to be extracted. A determination of the stable Cs-133 in this extract fraction supplies information about the final equilibrium state of caesium in the investigated soil at the moment of sampling.

Generally, the stable caesium concentrations in soils worldwide range from 1mg to 5mg per kg. After an extraction, the caesium concentration in the extract can be reduced by a thousand times, compared to the original soil sample. Determination of stable caesium in a soil-extract solution at this concentration level, however, presents a complex analytical problem. One of the methods sensitive enough at this concentration level is the large-sample activation analysis. This method is practically free from losses in the determination process and, apart from caesium concentration, gives information about other relevant elements present in the sample.

## 2. Experimental Methods

### 2.1. SAMPLES

Three different sample types were tested in this analysis: soils, wheat grain growing on these soils, and 1-M ammonium acetate soil extracts. The wheat grain samples with their host soils were collected in Austria, Denmark, Sweden, and Kirghyzistan. To test the conditions of large-sample irradiation, several seawater and brackish water samples and six different grain samples were also used. Soil samples collected from about top 5 cm were dried at 60° C. After drying, the debris and litter were manually removed. Grain samples were dried at 60° C and packed directly into irradiation containers without homogenisation. The ammonium acetate was analytical grade Merck, dissolved in triple-distilled water.

Soil samples were irradiated in the DR3 reactor in Denmark. Plant samples as well as water, brackish water, and ammonium oxalate soil extracts were irradiated in the Triga reactor at the Atominstitut in Vienna.

In the DR-3 reactor, irradiation of standards and soil samples was performed in a channel with a thermal flux of $4 \times 10^{13} \cdot cm^{-2} \cdot s^{-1}$. The irradiation time was 2 days and the sample weight ranged from 80 to 120 mg. All the samples were measured after 8 to 10 days decay time, and again after 20 to 30 days. Measuring time was 20 minutes and 4 hours for the first and the second measurements respectively. The measurement system consisted of a 51% high purity Ortec Detector, Ortec amplifier 673, Ortec Hi-Rate amplifier 973, spectrum buffer Spectrum Master 921, and an interface to a personal computer. Detector had a resolution of 1.75 keV for 1332 line. For acquisition the Ortec software was used. Area calculations for all the spectra were performed by means of the authors' AREA program, which is based on square-wave convolution.

In the Triga Mark II reactor of the Atominstitut in Vienna, large irradiation samples have been performed. The neutron flux in five dry irradiation holes is $1.8 \cdot 10^{12}$. Standards in 0.7-ml and 1.5-dram vials were used to calibrate the irradiation positions and the measuring system. The large samples were packed into the original General Dynamics hardened polyethylene containers. Because their useful volume is about 40 ml, two containers can be connected into one unit. All the solutions and organic samples were irradiated in such double units, with 40 ml of each sample in the bottom, leaving the upper unit empty. In the cover of each upper unit, a 5-mm hole was drilled. For samples with greater volume, 2, 3, 4 or 5 irradiation channels were used. After the 2-week decay time, the 40-ml (or 80-, 120-, 160-, or 200-ml) samples were manually mixed in a porcelain bowl (plants) or in 1 litre beakers (solutions) and repacked into 20 ml polyethylene bottles to be gamma counted in a radial holder [6].

The counting system consisted of a 15% Nuclear Data, high-purity germanium detector, an Ortec amplifier 673 in the gated integrator mode, an ND599 LFC unit, and the Accuspec B card. For shielding, 10 cm of iron were used. The resolution of the detector was 1.9 keV. For acquisition the Nuclear Data program was used. Area calculations for all the spectra were performed by means of the authors' AREA program and by the Peak Nuclear Data program.

In-situ temperature measurements inside the irradiated water, sea water, brackish water, and 1-M ammonium acetate containers were performed by means of a NiCr-Ni thermocouple insulated by a glass-fibre coating. The voltage was measured by means of a METEX Model M4650 CR voltmeter with a digital output to a Notebook computer in intervals of 5 s [7].

All the samples were measured in radial geometry by a special holder prepared for this purpose. A 1000-ml Marinelli beaker had its sides cut so that only a concentric ring was left around the bottom, holding up to thirteen 20 ml bottles. In this way, the collective volume of the thirteen subsamples could reach 200–260 ml in one simultaneous measurement [7].

## 3. Results and Discussion

As a sensitive multielement technique, the activation analysis was used in numerous analytical studies of soils, plants, and various soil extracts [8,9]. Before analysis often a preconcentration has to be applied, owing to a very low concentration of some essential elements. The most common preconcentration methods are either evaporation to a dry residue in case of an extractant solution, or ashing of a dried plant material. Usually only a minute amount of an inorganic residue or ash (ca 100 mg) in a hermetized quartz, plastic, or aluminium container becomes irradiated in a reactor channel. Many small samples are packed together with standards and an irradiation monitor into one irradiation container. The possibility of losses and contamination is an important drawback to all these preconcentration methods. In the case of a plant material and of a dry solution extract, an additional limiting factor consists of a very-low activity of the irradiated tiny samples. Generally in activation analysis, as in any other analytical method, the bigger the analytical signal the higher the sensitivity that can be achieved. An improvement in the signal can be achieved either by a stronger irradiation (a higher thermal flux is not always available) or by irradiating larger samples. Irradiation of large samples has an additional advantage in reducing or avoiding the contamination and losses connected to homogenisation and any other sample pretreatment before activation. Problems to be solved during an irradiation of large samples and working methods are described elsewhere [10].

### 3.1. TEMPERATURE MEASUREMENT

Irradiating techniques of a large volume of an aqueous solution constitutes the greatest difference compared to of the conventional irradiation techniques used in the neutron-activation analysis. The temperature increase measured during irradiation of many plants, organic substances, and metal oxides did not exceed 40° C. Only during an irradiation of samples with a high concentration of neutron-absorbing elements (such as boron), or of high Z elements (such as Pb), was the temperature measured to be from 100 to 120° C. Finding a high concentration of these two material types in an organic material, in plants, or in a water solution is quite improbable.

Since the only available data concerned irradiation of distilled water[7], an additional series of temperature measurements was performed for brackish water as a simulator of a 0.1 M NaCl (or KCl, or $CaCl_2$) extractant. Seawater from the Baltic Sea, four brackish water samples from the Danish fjords, and a 1-M ammonium acetate solution were irradiated for 6 hours in a dry channel of the Triga reactor. The results as presented in Figure 1 for a 1-M ammonium oxalate solution were very similar to those described for water [7]. After a fast rise in the beginning of an irradiation, the temperature rose very slowly to 38–39° C. For comparison's sake, a temperature curve for lentils is shown. Here the highest temperature attained was from 40 to 41° C. This high temperature resulted from the low thermal conductivity among many grain boundaries of single lentils. A lower temperature for water solutions is caused by better cooling conditions. Small differences (1 to 2 degrees) among various water solutions are caused by the varying contact between the plastic irradiation container and the walls of an irradiation tube. The diameter of an irradiation tube is 5 cm while the diameter of an irradiation container is 2.54 cm and 3.0 cm at the screwed lid part. There is no certainty that a plastic container would contact the bottom of a 6-m aluminium dry tube (reactor pool temperature is 29° C) in an exactly replicable way. Another source of variability of response is the position of a thermocouple in the irradiated sample. Its position can be slightly eccentric, which would shift the reading as compared to the reading from a strictly centred thermocouple.

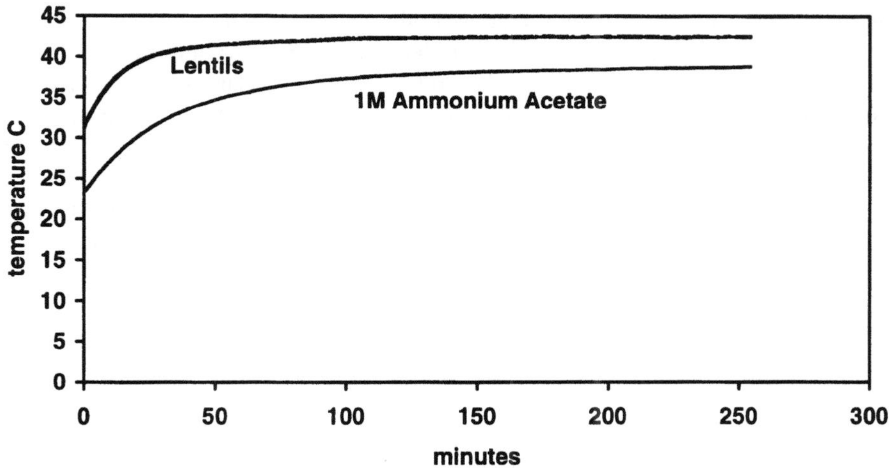


*Figure 1.* Temperature inside the irradiation container during irradiation.

## 3.2. SAMPLE WEIGHT

In the instrumental-activation analysis, the sensitivity of detection, or rather the detection limit for an element of interest, is very dependent on the sample composition. For geological samples and soil-related samples the main gamma emitters after an irradiation are sodium, lanthanum, iron, scandium, and cobalt. The activity of sodium, lanthanum and other short-lived isotopes, like arsenic, potassium, tungsten, and samarium, decays almost completely during the first 8 to 15 days after an irradiation. The activity of iron, scandium, cobalt, caesium, and some of the rare earths is longer living and hence constitutes a factor seriously limiting the weight of a soil-related sample. An irradiation of a sample exceeding 200–300 mg would produce an activity too high to be measured and too dangerous to be handled.

The situation is practically the opposite in the case of irradiation of organic materials, such as green plants, cereals and nuts, or water and water extracts. The main activity in this type of sample originates from short-lived potassium, sodium, bromine, and, sometimes, from arsenic and calcium. After an irradiation of the usual 100-mg amounts in hermetically sealed containers, the short-lived gamma-activity decays completely within the first 20 days and the remaining long-lived activity is very low. A better detection can be achieved either by ashing and evaporation or by a very high irradiation flux. Table 1 lists the results for the chemical composition of a soil sample irradiated in the DR-3 reactor in Denmark as well as the results of the corresponding 1-M ammonium acetate extract and wheat grain, irradiated as open large samples in the Triga reactor in Austria. The concentration is given in ng/g. Iron, cobalt and scandium, i.e., the elements producing hard gamma-emitting isotopes, are shown to have some hundred or thousand times higher concentration in a soil than in the corresponding ammonium acetate extract or wheat grain. In the absence of a high Compton continuum produced by hard gamma emitters, the large sample activation method gives an excellent sensitivity. Generally all the soil extracts and plant samples have a low concentration of microelements producing strong gamma-emitting isotopes; therefore large samples can be safely irradiated, and the detection sensitivity improves almost linearly with the growing weight of the sample. In brief, the larger and purer the sample, the better the sensitivity and the lower the detection limit. The first application of this principle was described in 1980 by Kim et al. [11]. An irradiation of 250 ml water in an open quartz container resulted in the detection of 19 elements on a level of from a picogram to a femtogram per gram.

TABLE 1. Concentration in ng/g in interrelated Danish samples

| Sample | Soil | Extract | Wheat grain |
|---|---|---|---|
| Sample weight | 80 mg | 160 g | 93 g |
| Na | 5740000 | 7850 | 9280 |
| Ba | 20800 | 2530 | 184 |
| Fe | 1903000 | 115 | 6230 |
| Zn | 2240 | 101 | 10000 |
| Br | 8080 | 90.7 | 480 |
| Ce | 2730 | 37.5 | 54.6 |
| Rb | 4400 | 17.1 | 2500 |
| La | 1200 | 5.22 | 0.56 |
| U | 109 | 3.36 | 7.1 |
| Sb | 38.1 | 3.19 | 1.22 |
| As | 7950 | 2.12 | 203 |
| Co | 249 | 1.64 | 2.19 |
| Cr | 2550 | 1.54 | 0 |
| Sm | 232 | 0.68 | 0 |
| Cs | 130 | 0.56 | 1.14 |
| Th | 337 | 0.45 | 1.71 |
| Eu | 35.1 | 0.17 | 0.03 |
| Sc | 5110 | 0.07 | 0.18 |

## 3.3. DISTRIBUTION COEFFICIENT

Many different extractants are used for isolation and extraction of toxic or radioactive contaminants from various soil phases. Elements in a soluble and exchangeable form are assumed to be transferred to plants. Ammonium acetate, sodium chloride, magnesium chloride, potassium, and sodium nitrate are the most popular neutral extractants used for assessment of the exchangeable part. The application and drawbacks of ammonium acetate use as a neutral extractant are extensively discussed by Del Castilho [12] and Ure [13]. The most serious limitation is the very low concentration of microelements in the liquid phase. For instance for a caesium there are numerous publications about the distribution coefficient of Cs-137, fewer for Cs-134, and only very few for Cs-133 [2-5]. The stable caesium isotope, which is present in nanogram or subnanogram quantities in a neutral extractant, is very difficult to determine. The information about the ageing effect on caesium availability in plants is based on radiometric measurements of Cs-134 and Cs-137 distribution.

The ageing effect exists as well for other elements, when radioactive tracers are added in experiments. Yasuda and Uchida [14] measured distribution coefficients for stable Mn, Zn, and Sr and for their radioactive tracers. In a hydrochloric acid solution, the distribution of radioactive species was from two to ten times higher than that for the stable ones. When the extractant consisted of deionized water, the extraction of radioactive species was even higher. Similarly the uptake of stable strontium by rice plants [15] was considerably lower than that of Sr-90.

In Table 2 the average results for four sandy soils from Denmark, Austria, Sweden, and Kirghyzistan are presented. The values in three columns are defined as follows:

$D_{SE}$ = concentration in soil / concentration in extract

$D_{PE}$ = concentration in plant / concentration in extract

$D_{PS}$ = concentration in plant / concentration in soil = $D_{PE}$ / $D_{SE}$

TABLE 2. Average distribution coefficients for four different soils

| | $D_{SE}$ | $D_{PE}$ | $D_{PS}$ |
|---|---|---|---|
| Zn | 4156 | 2220 | 0.534 |
| Br | 117 | 12.5 | 0.107 |
| Rb | 2230 | 66.2 | 0.0297 |
| As | 3750 | 96 | 0.0256 |
| Ba | 187 | 1.3 | 0.00695 |
| Ce | 3600 | 4.7 | 0.0013 |
| Sb | 980 | 0.9 | 0.00092 |
| Fe | 224000 | 157 | 0.0007 |
| Cs | 11970 | 3.6 | 0.0003 |
| Co | 10800 | 3.4 | 0.0003 |
| Th | 21900 | 4 | 0.00018 |
| La | 229 | 0.11 | 0.00048 |
| Cr | 30700 | 12.4 | 0.0004 |
| Eu | 23600 | 0.7 | 0.000029 |
| Sc | 712000 | 11.1 | 0.000016 |

Table 2 contains only these results, when all the three concentrations were determined in the soil/extract/plant triads. There were measured 15 such triads. In every sample, 29 microelements were looked for. Generally the detection limit in the ammonium acetate extract and in plants was 100 to 1000 times lower than the detection limit in soils. The elements undetected in a soil, but detected at a nanogram and picogram level in one or more extract or plant sample are the following : Se, Cr, Ag, Sr, Sm, Nd, Tb, Yb, Lu, Hf, Ta, Au, Hg, and U.

Because the measurement of distribution coefficients for radioactive species is very simple experimentally, the method is rather popular. The large-sample-activation analysis can be experimentally quite difficult, but it supplies accurate information about the simultaneous transfer of *stable elements* into plants. Wherever a radioactive soil contamination occurs or is simulated, or wherever leaching of radioactive waste is studied, the determination of the distribution of both the stable and radioactive isotopes of elements can be of essential interest.

## 4. Conclusions

The experimental results prove the determination of trace elements in plant samples and extracts simulating soil solutions to be possible on a low nanogram and picogram level. In these matrices such a low detection level is otherwise very difficult or impossible to achieve by means of any other analytical methods.

## 5. References

1. Evaluation of data on the transfer of radionuclides in the food chain. Post-Chernobyl action. Ed.by G. Desmet, J. Sinnaeve in EUR 12550, 1992
2. A. Wallace, E M Romney, R A Wood, Soil Science, 134(1982), 1, 71-75
3. P. Varskov, R. Naeumann, E. Steinnes, J. Environ. Radioactivity, 22(1994), 43-53
4. B. Salbu, E. Steinnes, Analyst, 117(1992), 243-249
5. A. Wyttenbach, V. Furrer, L. Tobler, Sci. Total Environm., 173/174(1995), 361-367
6. R. Gwozdz, F. Grass, in Environmental Radiochemical Analysis, 289-300, Ed. G W A Newton, the Royal Society of Chemistry, Cambridge, UK, 1999
7. J. Casta, R. Gwozdz, MTAA10, Bethesda, MD 19-23 April, 1999, in press
8. R. A. Nadkarni, Radiochem. Radioanal. Letters 26(1976), 2, 139-150
9. S. Bajo, A. Wyttenbach, L. Tobler, H. Conradin, J. Radioanal. Nucl. Chem. 134(1989) 1, 181 –191
10. R. Gwozdz, F. Grass, MTAA10, Bethesda, MD, 19-23 April, 1999, in press
11. J. Kim, G. Bohm, R. Henkelman, Fresen. Z. Anal. Chem., 327(1975), 153
12. P. Del Castilho, Intern. J. Environ. Anal. Chem., 51(1993), 59-64
13. A.M Ure, Sci. Total Environment, 178(1996), 3-10
14. H. Yasuda, S. Ushida, J. Health Phys., 30(1991), 15-18
15. M. Komamura, A. Tsumura, Radioisotopes, 43(1994), 1

## LIST OF AUTHORS

S.R. ADAMS
*2621 N. Losee Rd.*
*Bechtel Nevada,*
*North Las Vegas, NV, 89030, USA*

M.A. AKHMETOV
*Institute of Radiation Safety and Ecology*
*National Nuclear Centre of the Republic of Kazakhstan*
*Kurchatov-City, 490021, Republic of Kazakhstan*

L.R. ANSPAUGH
*Radiobiology Division*
*University of Utah*
*729 Arapeen Drive*
*Salt Lake City, UT 84108-1218, USA*

O.I. ARTEMEV
*Institute of Radiation Safety and Ecology*
*National Nuclear Centre of the Republic of Kazakhstan*
*Kurchatov-City, 490021, Republic of Kazakhstan*

I.B. ASTAKHOV
*Russian Federal Nuclear Centre (RFNC),*
*All-Russian Scientific Institute of Experimental Physics (VNIIEF),*
*Arzamas-16, Sarov, Russia*

D.N. BELYASHOV
*Institute of Geophysical Investigations,*
*National Nuclear Centre of the Republic of Kazakhstan,*
*Altmay, 480082, Republic of Kazakhstan*

P.V. CHAKROV
*Institute of Nuclear Physics,*
*National Nuclear Centre of the Republic of Kazakhstan,*
*Almaty, 480082, Republic of Kazakhstan*

G.N. CHUMIKOV
*Institute of Nuclear Physics,*
*National Nuclear Centre of the Republic of Kazakhstan,*
*Almaty, 480082, Republic of Kazakhstan*

B.W. CHURCH
*BWC Enterprises, Inc.,*
*1795 Whipple Ave.*
*Logandale, NV, 89021, USA*

J. CONCA
*Environmental Technology Division,*
*Los Alamos National Laboratory,*
*Los Alamos, NM, 87545, USA*

P .R. DANESI
*International Atomic Energy Agency,*
*IAEA Laboratories-Seibersdorf,*
*P.O. Box 100, A-1400, Vienna , Austria*

D.K. DAUKEEV
*National Nuclear Centre of the Republic of Kazakhstan,*
*Almaty, 480082, Republic of Kazakhstan*

A.A. DRUZHININ
*Russian Federal Nuclear Centre (RFNC),*
*All-Russian Scientific Institute of Experimental Physics (VNIIEF),*
*Arzamas-16, Sarov, Russia*

D.W. EFURD
*Nuclear and Radiochemistry Division,*
*Los Alamos National Laboratory,*
*Los Alamos, NM, 87545, USA*

A.ELISEEV
*Institute of Nuclear Physics,*
*National Nuclear Centre of the Republic of Kazakhstan,*
*Almaty, 480082, Republic of Kazakhstan*

D. L. FINNEGAN
*Chemical Science and Technology Division,*
*Los Alamos National Laboratory,*
*Los Alamos, NM 87545, USA*

F. GRASS
*Atominstitut,*
*Schuettelstr. 115, 1020 Vienna, Austria*

YU.A. GRINSHTEIN
*Institute of Geophysical Investigations,*
*National Nuclear Centre of the Republic of Kazakhstan,*
*Altmay, 480082, Republic of Kazakhstan*

R.GWOZDZ
*Tracechem,*
*Markmandsgade 2, 3 th, 2300 Copenhagen S, Denmark*

M.J. HAMEEDI
*National Oceanic and Atmospheric Administration,*
*Silver Spring, MD, 20910, USA*

W. L. HAWKINS
*Earth and Environmental Science Division,*
*Los Alamos National Laboratory,*
*Los Alamos, NM 87545, USA*

S. S. HECKER
*Senior Fellow, Los Alamos National Laboratory,*
*Los Alamos, NM 87545, USA*

R. W. HENDERSON
*Environment, Safety, and Health Division,*
*Los Alamos National Laboratory,*
*Los Alamos, NM 87545, USA*

R.I. ILKAYEV
*Russian Federal Nuclear Centre (RFNC)*
*All-Russian Scientific Institute of Experimental Physics (VNIIEF)*
*Arzamas-16, Sarov, Russia*

W.C.T. INKRET
*Radiological Dose Assessment,*
*Los Alamos National Laboratory,*
*Los Alamos, NM, 87545, USA*

K.K. KADYRZHANOV
*Institute of Nuclear Physics,*
*National Nuclear Centre of the Republic of Kazakhstan,*
*Almaty, 480082, Republic of Kazakhstan*

I.V. KAZACHEVSKY
*Institute of Nuclear Physics,*
*National Nuclear Centre of the Republic of Kazakhstan,*
*Almaty, 480082, Republic of Kazakhstan*

I.V. KAZACHEVSKIY
*Institute of Nuclear Physics,*
*National Nuclear Centre of the Republic of Kazakhstan,*
*Almaty, 480082, Republic of Kazakhstan*

I.M. KHAIKOVICH
*Research Institute of Prospecting Geophysics (RIPG),*
*St. Petersburg, Russia*

S. KHAZHEKBER
*Institute of Nuclear Physics,*
*National Nuclear Centre of the Republic of Kazakhstan,*
*Almaty, 480082, Republic of Kazakhstan*

I.V. KHROMUSHIN
*National Nuclear Centre of the Republic of Kazakhstan,*
*Almaty, 480082, Republic of Kazakhstan*

K.M. KIKKARIN
*Institute of Nuclear Physics,*
*National Nuclear Centre of the Republic of Kazakhstan,*
*Almaty, 480082, Republic of Kazakhstan*

S.B. KISLITSIN
*Institute of Nuclear Physics,*
*National Nuclear Centre of the Republic of Kazakhstan,*
*Almaty, 480082, Republic of Kazakhstan*

A.M. KOROCHKIN
*Russian Federal Nuclear Centre (RFNC),*
*All-Russian Scientific Institute of Experimental Physics (VNIIEF),*
*Arzamas-16, Sarov, Russia*

N.G. KRYLOV
*Russian Federal Nuclear Centre (RFNC),*
*All-Russian Scientific Institute of Experimental Physics (VNIIEF),*
*Arzamas-16, Sarov, Russia*

A.T. KUKETAEV
*Institute of Nuclear Physics,*
*National Nuclear Centre of the Republic of Kazakhstan,*
*Almaty, 480082, Republic of Kazakhstan*

YU. KUTLAKHMEDOV
*Biological Faculty,*
*Kiev University,*
*Vladimirskaya Str. 64, 252601 Kiev, Ukraine*

V.N. LARIN
*Institute of Radiation Safety and Ecology*
*National Nuclear Centre of the Republic of Kazakhstan,*
*Kurchatov-City, 490021, Republic of Kazakhstan*

A.D. LOPUGA
*Institute of Nuclear Physics,*
*National Nuclear Centre of the Republic of Kazakhstan,*
*Almaty, 480082, Republic of Kazakhstan*

N. LU
*Environmental Technology Division,*
*Los Alamos National Laboratory,*
*Los Alamos, NM, 87545, USA*

S.N. LUKASHENKO
*Institute of Nuclear Physics,*
*National Nuclear Centre of the Republic of Kazakhstan,*
*Almaty, 480082, Republic of Kazakhstan*

M. MAKAREWICZ
*International Atomic Energy Agency,*
*IAEA Laboratories-Seibersdorf,*
*P.O. Box 100, A-1400, Vienna, Austria*

M. YU. MAKSIMOV
*Russian Federal Nuclear Centre (RFNC),*
*All-Russian Scientific Institute of Experimental Physics (VNIIEF),*
*Arzamas-16, Sarov, Russia*

A. MARKOWICZ
*International Atomic Energy Agency,*
*IAEA Laboratories-Seibersdorf,*
*P.O. Box 100, A-1400, Vienna, Austria*

L.J. MARTIN
*Australian Radiation Protection and Nuclear Safety Agency,*
*Yallambie, Victoria, Australia*

C.F.V. MASON
*Nuclear Materials Technology Division,*
*Los Alamos National Laboratory,*
*Los Alamos, NM, 87545, USA*

M. MATTIUZZI
*International Atomic Energy Agency,*
*IAEA Laboratories-Seibersdorf,*
*P.O. Box 100, A-1400, Vienna, Austria*

J. MORENO
*International Atomic Energy Agency,*
*IAEA Laboratories-Seibersdorf,*
*P.O. Box 100, A-1400, Vienna, Austria*

L.M. NAZARENKO
*Institute of Nuclear Physics,*
*National Nuclear Centre of the Republic of Kazakhstan,*
*Almaty, 480082, Republic of Kazakhstan*

R.S. O'BRIEN
*Australian Radiation Protection and Nuclear Safety Agency,*
*Yallambie, Victoria, Australia*

P. ORMAI
*Public Agency for Radioactive Waste Management (PURAM),*
*H-2040 Budaörs, Puskas Tivadar str. 11, Hungary*

D. OUGHTON
*Department of Chemistry and Biotechnology,*
*P.O. Box 5026, Agricultural University of Norway,*
*1432 Aas, Norway*

S.P. PIVOVAROV
*Institute of Nuclear Physics,*
*National Nuclear Centre of the Republic of Kazakhstan,*
*Almaty, 480082, Republic of Kazakhstan*

A.PLATOV
*Institute of Nuclear Physics,*
*National Nuclear Centre of the Republic of Kazakhstan,*
*Almaty, 480082, Republic of Kazakhstan*

G. POLIKARPOV
*Faculty of Comparative Radioecology*
*Institute of Biology of the Southern Seas,*
*Prosp. Nachimova, 2, 335011 Sevastopol, Ukraine*

L.D. PTITSKAYA
*Institute of Radiation Safety and Ecology*
*National Nuclear Centre of the Republic of Kazakhstan,*
*Kurchatov-City, 490021, Republic of Kazakhstan*

P.DE REGGE
*International Atomic Energy Agency,*
*IAEA Laboratories-Seibersdorf,*
*P.O. Box 100, A-1400, Vienna, Austria*

F.R. ROENSCH
*Nuclear and Radiochemistry Division,*
*Los Alamos National Laboratory,*
*Los Alamos, NM, 87545, USA*

A.B. RUKHIN
*Institute of Nuclear Physics,*
*National Nuclear Centre of the Republic of Kazakhstan,*
*Almaty, 480082, Republic of Kazakhstan*

B. SALBU
*Department of Chemistry and Biotechnology,*
*P.O. Box 5026, Agricultural University of Norway,*
*1432 Aas, Norway*

B. SCHEBER
*Center for Army Analysis,*
*Attn: CSCA-EN,*
*6001 Goethals Rd., Suite 102,*
*Fort Belvoir, VA, 22060, USA*

F. SCHOENHOFER
*International Atomic Energy Agency,*
*IAEA Laboratories-Seibersdorf,*
*P.O. Box 100, A-1400, Vienna , Austria*

T.A. SEREDAVINA
*Institute of Nuclear Physics,*
*National Nuclear Centre of the Republic of Kazakhstan,*
*Almaty, 480082, Republic of Kazakhstan*

J.H. SHINN
*Lawrence Livermore National Laboratory,*
*Livermore, CA, 94551, USA*

S. L. SIMON
*National Cancer Institute,*
*6120 Executive Blvd., MSC 7238*
*Executive Plaza South,*
*Bethesda, MD, 20852, USA*

G.P. SILKINA
*Institute of Radiation Safety and Ecology*
*National Nuclear Centre of the Republic of Kazakhstan,*
*Kurchatov-City, 490021, Republic of Kazakhstan*

KH. SIRAJET
*Institute of Nuclear Physics,*
*National Nuclear Centre of the Republic of Kazakhstan,*
*Almaty, 480082, Republic of Kazakhstan*

L. SKIPPERUD
*Department of Chemistry and Biotechnology,*
*P.O. Box 5026, Agricultural University of Norway,*
*1432 Aas, Norway*

O. SLÁVIK
*VU'JE Research Institute,*
*Trnava, Inc.,*
*Okruzna 5*
*917 00, Trnava, Slovakia*

R. F. SMALE
*Environment, Safety, and Health Division,*
*Los Alamos National Laboratory,*
*Los Alamos, NM 87545, USA*

V.P. SOLODUKHIN
*Institute of Nuclear Physics,*
*National Nuclear Centre of the Republic of Kazakhstan,*
*Almaty, 480082, Republic of Kazakhstan*

J. SOLYMOSI
*Miklos Zrinyi National Defence University,*
*Department of Environmental and NBC Security,*
*H-1581 Budapest P.O.B. 15, Hungary*

S.B. SUBBOTIN
*Institute of Radiation Safety and Ecology*
*National Nuclear Centre of the Republic of Kazakhstan,*
*Kurchatov-City, 490021, Republic of Kazakhstan*

A. TAJANI
*International Atomic Energy Agency,*
*IAEA Laboratories-Seibersdorf,*
*P.O. Box 100, A-1400, Vienna, Austria*

J. L. THOMPSON
*Chemical Science and Technology Division,*
*Los Alamos National Laboratory,*
*Los Alamos, NM 87545, USA*

T.E. TURKEBAEV
*Institute of Nuclear Physics,*
*National Nuclear Centre of the Republic of Kazakhstan,*
*Almaty, 480082, Republic of Kazakhstan*

M.F. VERESTCHAK
*Institute of Nuclear Physics,*
*National Nuclear Centre of the Republic of Kazakhstan,*
*Almaty, 480082, Republic of Kazakhstan*

Á. VINCZE
*Miklos Zrinyi National Defence University,*
*Department of Environmental and NBC Security,*
*H-1581 Budapest P.O.B. 15, Hungary*

G. VOLENT
*Paks Nuclear Power Plant Ltd.,*
*Paks,*
*P.O. Box 71, H–7031 Hungary*

G.A. WILLIAMS
*Australian Radiation Protection and Nuclear Safety Agency,*
*Yallambie, Victoria, Australia*

E.M. YAKUSHEV
*Institute of Nuclear Physics,*
*National Nuclear Centre of the Republic of Kazakhstan,*
*Almaty, 480082, Republic of Kazakhstan*

E. ZEILLER
*International Atomic Energy Agency,*
*IAEA Laboratories-Seibersdorf,*
*P.O. Box 100, A-1400, Vienna, Austria*

A.K. ZHETBAEV
*Institute of Nuclear Physics,*
*National Nuclear Centre of the Republic of Kazakhstan,*
*Almaty, 480082, Republic of Kazakhstan*

K. ZHUMADILOV
*Institute of Radiation Safety and Ecology*
*National Nuclear Centre of the Republic of Kazakhstan,*
*Kurchatov-City, 490021, Republic of Kazakhstan*

MIX
Papier aus verantwortungsvollen Quellen
Paper from responsible sources
FSC® C105338

If you have any concerns about our products,
you can contact us on
**ProductSafety@springernature.com**

In case Publisher is established outside the EU,
the EU authorized representative is:
**Springer Nature Customer Service Center GmbH**
**Europaplatz 3, 69115 Heidelberg, Germany**

Printed by Libri Plureos GmbH
in Hamburg, Germany